Marine Ecology

Marine Ecology

Edited by
Rafe Shaw

Larsen & Keller
www.larsen-keller.com

Marine Ecology
Edited by Rafe Shaw
ISBN: 978-1-63549-173-9 (Hardback)

© 2017 Larsen & Keller

$\blacksquare$ Larsen & Keller

Published by Larsen and Keller Education,
5 Penn Plaza,
19th Floor,
New York, NY 10001, USA

Cataloging-in-Publication Data

Marine ecology / edited by Rafe Shaw.
    p. cm.
Includes bibliographical references and index.
ISBN 978-1-63549-173-9
1. Marine ecology. 2. Aquatic ecology. 3. Ecology. I. Shaw, Rafe.
QH541.5.S3 M36 2017
577.7--dc23

The publisher's policy is to use permanent paper from mills that operate a sustainable forestry policy. Furthermore, the publisher ensures that the text paper and cover boards used have met acceptable environmental accreditation standards.

Printed and bound in the United States of America.

For more information regarding Larsen and Keller Education and its products, please visit the publisher's website www.larsen-keller.com

Table of Contents

Permissions

Index

Preface

This book is a compilation of chapters that discuss the most vital concepts in the field of marine ecology. It gives thorough insights about this vast field and provides detailed information about the various advances in this area. Marine ecology is the study of the different marine organisms, their environment, population and their interactions with each other and their surroundings. This textbook is complied in such a manner that it will provide in-depth knowledge about the theory and practice of marine ecology. The topics covered in this extensive book deal with the core subjects of marine ecology. Those in search of information to further their knowledge will be greatly assisted by this text.

A detailed account of the significant topics covered in this book is provided below:

Chapter 1 - Marine ecology is very important for the well-being of our planet. Marine ecology is very diverse and covers two-thirds of the surface of the Earth. The chapter on marine ecology offers an insightful focus, keeping in mind the subject matter.

Chapter 2 - The various components of marine ecosystem that have been discussed in this section are salt marsh, estuary, intertidal zone, lagoon, deep sea and benthos. The lowest layer found in any ocean is known as deep sea whereas benthos are organisms that live on or near the seabed. The section strategically encompasses and incorporates the major components and key concepts of marine ecosystem, providing a complete understanding.

Chapter 3 - The protection and conservation of marine ecosystems is termed as marine conservation. It aims to limit the harm caused by humans on the marine ecosystem as well as restoring damaged ecosystems. The aspects elucidated in this text are of vital importance, and provide a better understanding of marine ecosystems.

Chapter 4 - The ecosystem that is found in a body of water is known as aquatic ecosystem. The two kinds of aquatic ecosystems are marine and freshwater ecosystem. Aquatic ecosystems also include freshwater ecosystem, lake ecosystem and river ecosystem. This section provides the reader with an integrated understanding of aquatic ecosystems.

Chapter 5 - Marine biology is the study of all the organisms that are found in the ocean. There is a difference between marine biology and marine ecology; the difference being is that marine biology studies the interaction between organisms whereas marine ecology studies the interaction between the environment and the organisms. This section helps the readers in understanding marine biology in detail.

Chapter 6 - To have a precise understanding of marine toxicology, it is very important to understand aquatic toxicology and cyanotoxin. Industrial waste and manufactured chemicals have an immense effect on marine biology. The study of these effects is known as aquatic toxicology. This section has been carefully written to provide an easy understanding of marine toxicology.

Chapter 7 - Marine life signifies the life of all the organisms and plants that live in the ocean. The area inhabited by marine species is known as marine habitat. Marine habitats are divided into coastal habitats and open ocean habitats. The topics covered in this chapter are very important in developing a complete understanding of the subject.

Editor

1

Introduction to Marine Ecology

Marine ecology is very important for the well-being of our planet. Marine ecology is very diverse and covers two-thirds of the surface of the Earth. The chapter on marine ecology offers an insightful focus, keeping in mind the subject matter.

Marine Ecosystem

Marine ecosystems are among the largest of Earth's aquatic ecosystems, salt marshes, intertidal zones, estuaries, lagoons, mangroves, coral reefs, the deep sea, and the sea floor. They can be contrasted with freshwater ecosystems, which have a lower salt content. Marine waters cover two-thirds of the surface of the Earth. Such places are considered ecosystems because the plant life supports the animal life and vice versa.

Marine ecosystems are very important for the overall health of both marine and terrestrial environments. According to the World Resource Center, coastal habitats alone account for approximately 1/3 of all marine biological productivity, and estuarine ecosystems (i.e., salt marshes, seagrasses, mangrove forests) are among the most productive regions on the planet. In addition, other marine ecosystems such as coral reefs, provide food and shelter to the highest levels of marine diversity in the world.

Marine ecosystems usually have a large biodiversity and are therefore thought to have a good resistance against invasive species. However, exceptions have been observed, and the mechanisms responsible in determining the success of an invasion are not yet clear.

Large Marine Ecosystem

Large marine ecosystems (LMEs) are regions of the world's oceans, encompassing coastal areas from river basins and estuaries to the seaward boundaries of continental shelves and the outer margins of the major ocean current systems. They are relatively large regions on the order of 200,000 km² or greater, characterized by distinct bathymetry, hydrography, productivity, and trophically dependent populations.

General characteristics of a large marine ecosystem (Gulf of Alaska)

The system of LMEs has been developed by the US National Oceanic and Atmospheric Administration (NOAA) to identify areas of the oceans for conservation purposes. The objective is to use the LME concept as a tool for enabling ecosystem-based management to provide a collaborative approach to management of resources within ecologically-bounded transnational areas. This will be done in an international context and consistent with customary international law as reflected in 1982 UN Convention on the Law of the Sea.

LME-based conservation is based on recognition that the world's coastal ocean waters are degraded by unsustainable fishing practices, habitat degradation, eutrophication, toxic pollution, aerosol contamination, and emerging diseases, and that positive actions to mitigate these threats require coordinated actions by governments and civil society to recover depleted fish populations, restore degraded habitats and reduce coastal pollution.

Although the LMEs cover only the continental margins and not the deep oceans and oceanic islands, the 64 LMEs produce 95% of the world's annual marine fishery biomass yields. Most of the global ocean pollution, overexploitation, and coastal habitat alteration occur within their waters. NOAA has conducted studies of principal driving forces affecting changes in biomass yields for 33 of the 64 LMEs, which have been peer-reviewed and published in ten volumes.

Large Marine Ecosystems (NOAA)

Global map of large marine ecosystems. Oceanographers and biologists have identified 64 LMEs worldwide.

- East Bering Sea

- Gulf of Alaska

- California Current

- Gulf of California

- Gulf of Mexico

- Southeast U.S. Continental Shelf

- Northeast U.S. Continental Shelf

- Scotian Shelf

- Newfoundland-Labrador Shelf

- Insular Pacific-Hawaiian

- Pacific Central-American Coastal

- Caribbean Sea

- Humboldt Current

- Patagonian Shelf

- South Brazil Shelf

- East Brazil Shelf

- North Brazil Shelf

- West Greenland Shelf

- East Greenland Shelf

- Barents Sea

- Norwegian Shelf

- North Sea

- Baltic Sea

- Celtic-Biscay Shelf

- Iberian Coastal

- Mediterranean Sea

- Canary Current

- Guinea Current
- Benguela Current
- Agulhas Current
- Somali Coastal Current
- Arabian Sea
- Red Sea
- Bay of Bengal
- Gulf of Thailand
- South China Sea
- Sulu-Celebes Sea
- Indonesian Sea
- North Australian Shelf
- Northeast Australian Shelf/Great Barrier Reef
- East-Central Australian Shelf
- Southeast Australian Shelf
- Southwest Australian Shelf
- West-Central Australian Shelf
- Northwest Australian Shelf
- New Zealand Shelf
- East China Sea
- Yellow Sea
- Kuroshio Current
- Sea of Japan
- Oyashio Current
- Sea of Okhotsk
- West Bering Sea
- Chukchi Sea

- Beaufort Sea

- East Siberian Sea

- Laptev Sea

- Kara Sea

- Iceland Shelf

- Faroe Plateau

- Antarctica

- Black Sea

- Hudson Bay

- Arctic Ocean

Hydrobiology

Hydrobiology is the science of life and life processes in water. Much of modern hydrobiology can be viewed as a sub-discipline of ecology but the sphere of hydrobiology includes taxonomy, economic biology, industrial biology, morphology, physiology etc. The one distinguishing aspect is that all relate to aquatic organisms. Much work is closely related to limnology and can be divided into lotic system ecology (flowing waters) and lentic system ecology (still waters).

One of the significant areas of current research is eutrophication. Special attention is paid to biotic interactions in plankton assemblage including the microbial loop, the mechanism of influencing water blooms, phosphorus load and lake turnover. Another subject of research is the acidification of mountain lakes. Long-term studies are carried out on changes in the ionic composition of the water of rivers, lakes and reservoirs in connection with acid rain and fertilisation. One goal of current research is elucidation of the basic environmental functions of the ecosystem in reservoirs, which are important for water quality management and water supply.

Much of the early work of hydrobiologists concentrated on the biological processes utilised in sewage treatment and water purification especially slow sand filters. Other historically important work sought to provide biotic indices for classifying waters according to the biotic communities that they supported. This work continues to this day in Europe in the development of classification tools for assessing water bodies for the EU water framework directive.

Field of Research Interests

The following are the research interests of hydrobiologists:

- acidification impact on lake and reservoir ecosystems
- ocean acidification
- paleolimnology of remote mountain lakes
- molecular ecology, phylogeography and taxonomy of Cladocera
- ultramorphology of cladoceran limbs and feeding adaptations
- chemical communication in plankton (prey-predator interaction)
- biomanipulation of water reservoirs
- cyclus of major nutrients (phosphorus, nitrogen)
- self-controlling mechanisms at population and community level

Organizations

- American Society of Limnology and Oceanography (ASLO)
- International Association of Theoretical and Applied Limnology (SIL)
- American Fisheries Society
- Freshwater Biological Association, England
- Marine Biological Laboratory (USA)
- Australian Society for Fish Biology
- Fisheries and Marine Institute of Memorial University of Newfoundland
- Department of Hydrobiology (Charles University, Prague)
- Dresden University of Technology Institute of Hydrobiology
- Institute of Hydrobiology and Fishery Science
- Water Research Institute T.G.M.
- Hydrobiological Institute, Academy of Science of Czech Republic
- Research Institute of Fish Culture and Hydrobiology
- Department of Hydrobiology, Slovak Academy of Science, Bratislava, Slovakia

- Max-Planck-Institut fur Limnologie in Ploen, Germany

- CNR-Istituto Italiano di Idrobiologia

- Institute of Hydrobiology, Chinese Academy of Sciences

- Hydrobiology Pty Ltd Brisbane, Australia based private consulting company

- Institute of Zoology and Hydrobiology University of Tartu, Estonia

- Department of Hydrobiology Bulgarian Academy of Sciences

- Department of General and Applied Hydrobiology Faculty of Biology, Sofia University "Sveti Kliment Ohridski", Bulgaria

References

- E.P.H. Best (Editor), Jan P. Bakker (Editor) Netherlands-Wetlands (Developments in Hydrobiology series) (Kluwer Academic Publishers, Dordrecht, 1993, 328 pp) ISBN 0-7923-2473-0

- R.I. Jones (Editor), V. Ilmavirta (Editor) Flagellates in Freshwater Ecosystems (Developments in Hydrobiology series) (Kluwer Academic Publishers, Dordrecht, 1992, 498pp.) ISBN 90-6193-651-9

- Jürgen Schwoerbel Methods of hydrobiology (freshwater biology) (Pergamon Press; [1st English ed.] edition, 1970, 200pp.) ISBN 0-08-006604-6

2

Various Components of Marine Ecosystem

The various components of marine ecosystem that have been discussed in this section are salt marsh, estuary, intertidal zone, lagoon, deep sea and benthos. The lowest layer found in any ocean is known as deep sea whereas benthos are organisms that live on or near the seabed. The section strategically encompasses and incorporates the major components and key concepts of marine ecosystem, providing a complete understanding.

Salt Marsh

A salt marsh or saltmarsh, also known as a coastal salt marsh or a tidal marsh, is a coastal ecosystem in the upper coastal intertidal zone between land and open salt water or brackish water that is regularly flooded by the tides. It is dominated by dense stands of salt-tolerant plants such as herbs, grasses, or low shrubs. These plants are terrestrial in origin and are essential to the stability of the salt marsh in trapping and binding sediments. Salt marshes play a large role in the aquatic food web and the delivery of nutrients to coastal waters. They also support terrestrial animals and provide coastal protection.

Basic Information

An estuarine salt marsh along the Heathcote River, Christchurch, New Zealand

Salt marshes occur on low-energy shorelines in temperate and high-latitudes which can be stable or emerging, or submerging if the sedimentation rate exceeds the subsidence rate. Commonly these shorelines consist of mud or sand flats (known also as tidal flats or abbreviated to mudflats) which are nourished with sediment from inflowing rivers and streams. These typically include sheltered environments such as embankments, estuaries and the leeward side of barrier islands and spits. In the tropics and sub-tropics they are replaced by mangroves; an area that differs from a salt marsh in that instead of herbaceous plants, they are dominated by salt-tolerant trees.

Salt marsh on Sapelo Island, Georgia, USA

Most salt marshes have a low topography with low elevations but a vast wide area, making them hugely popular for human populations. Salt marshes are located among different landforms based on their physical and geomorphological settings. Such marsh landforms include deltaic marshes, estuarine, back-barrier, open coast, embayments and drowned-valley marshes. Deltaic marshes are associated with large rivers where many occur in Southern Europe such as the Camargue, France in the Rhone delta or the Ebro delta in Spain. They are also extensive within the rivers of the Mississippi Delta in the United States. In New Zealand, most salt marshes occur at the head of estuaries in areas where there is little wave action and high sedimentation. Such marshes are located in Awhitu Regional Park in Auckland, the Manawatu Estuary, and the Avon-Heathcote Estuary in Christchurch. Back-barrier marshes are sensitive to the reshaping of barriers in the landward side of which they have been formed. They are common along much of the eastern coast of the United States and the Frisian Islands. Large, shallow coastal embayments can hold salt marshes with examples including Morecambe Bay and Portsmouth in Britain and the Bay of Fundy in North America.

Salt marshes are sometimes included in lagoons, and the difference is not very marked; the Venetian Lagoon in Italy, for example, is made up of these sorts of animals and or living organisms belonging to this ecosystem. They have a big impact on the biodiversity of the area. Salt marsh ecology involves complex food webs which include primary producers (vascular plants, macroalgae, diatoms, epiphytes, and phytoplankton), primary consumers (zooplankton, macrozoa, molluscs, insects), and secondary consumers.

The low physical energy and high grasses provide a refuge for animals. Many marine fish use salt marshes as nursery grounds for their young before they move to open waters. Birds may raise their young among the high grasses, because the marsh provides both sanctuary from predators and abundant food sources which include fish trapped in pools, insects, shellfish, and worms.

Formation

The formation begins as tidal flats gain elevation relative to sea level by sediment accretion, and subsequently the rate and duration of tidal flooding decreases so that vegetation can colonize on the exposed surface. The arrival of propagules of pioneer species such as seeds or rhizome portions are combined with the development of suitable conditions for their germination and establishment in the process of colonisation. When rivers and streams arrive at the low gradient of the tidal flats, the discharge rate reduces and suspended sediment settles onto the tidal flat surface, helped by the backwater effect of the rising tide. Mats of filamentous blue-green algae can fix silt and clay sized sediment particles to their sticky sheaths on contact which can also increase the erosion resistance of the sediments. This assists the process of sediment accretion to allow colonising species (e.g., Salicornia spp.) to grow. These species retain sediment washed in from the rising tide around their stems and leaves and form low muddy mounds which eventually coalesce to form depositional terraces, whose upward growth is aided by a sub-surface root network which binds the sediment. Once vegetation is established on depositional terraces further sediment trapping and accretion can allow rapid upward growth of the marsh surface such that there is an associated rapid decrease in the depth and duration of tidal flooding. As a result, competitive species that prefer higher elevations relative to sea level can inhabit the area and often a succession of plant communities develops.

Tidal Flooding and Vegetation Zonation

An Atlantic coastal salt marsh in Connecticut.

Coastal salt marshes can be distinguished from terrestrial habitats by the daily tidal flow that occurs and continuously floods the area. It is an important process in delivering sediments, nutrients and plant water supply to the marsh. At higher elevations in the upper marsh zone, there is much less tidal inflow, resulting in lower salinity levels. Soil salinity in the lower marsh zone is fairly constant due to everyday annual tidal flow. However, in the upper marsh, variability in salinity is shown as a result of less frequent flooding and climate variations. Rainfall can reduce salinity and evapotranspiration can increase levels during dry periods. As a result, there are microhabitats populated by different species of flora and fauna dependant on their physioglical abilities. The flora of a salt marsh is differentiated into levels according to the plants' individual tolerance of salinity and water table levels. Vegetation found at the water must be able to survive high salt concentrations, periodical submersion, and a certain amount of water movement, while plants further inland in the marsh can sometimes experience dry, low-nutrient conditions. It has been found that the upper marsh zones limit species through competition and the lack of habitat protection, while lower marsh zones are determined through the ability of plants to tolerate physiological stresses such as salinity, water submergence and low oxygen levels.

High marsh in the Marine Park Salt Marsh Nature Center in Brooklyn, New York

The New England salt marsh is subject to strong tidal influences and shows distinct patterns of zonation. In low marsh areas with high tidal flooding, a monoculture of the smooth cordgrass, *Spartina alterniflora* dominate, then heading landwards, zones of the salt hay, *Spartina patens*, black rush, *Juncus gerardii* and the shrub *Iva frutescens* are seen respectively. These species all have different tolerances that make the different zones along the marsh best suited for each individual.

Plant species diversity is relatively low, since the flora must be tolerant of salt, complete or partial submersion, and anoxic mud substrate. The most common salt marsh plants are glassworts (*Salicornia* spp.) and the cordgrass (*Spartina* spp.), which have worldwide distribution. They are often the first plants to take hold in a mudflat and begin its ecological succession into a salt marsh. Their shoots lift the main flow of the tide above the mud surface while their roots spread into the substrate and stabilize the sticky mud and carry oxygen into it so that other plants can establish themselves as well. Plants such as sea lavenders (*Limonium* spp.), plantains (*Plantago* spp.), and varied sedges and rushes grow once the mud has been vegetated by the pioneer species.

Salt marshes are quite photosynthetically active and are extremely productive habitats. They serve as depositories for a large amount of organic matter and are full of decomposition, which feeds a broad food chain of organisms from bacteria to mammals. Many of the halophytic plants such as cordgrass are not grazed at all by higher animals but die off and decompose to become food for micro-organisms, which in turn become food for fish and birds.

Sediment Trapping, Accretion, and the Role of Tidal Creeks

Bloody Marsh in Georgia, USA

The factors and processes that influence the rate and spatial distribution of sediment accretion within the salt marsh are numerous. Sediment deposition can occur when marsh species provide a surface for the sediment to adhere to, followed by deposition onto the marsh surface when the sediment flakes off at low tide. The amount of sediment adhering to salt marsh species is dependent on the type of marsh species, the proximity of the species to the sediment supply, the amount of plant biomass, and the elevation of the species. For example, in a study of the Eastern Chongming Island and Jiuduansha Island tidal marshes at the mouth of the Yangtze River, China, the amount of sediment adhering to the species *Spartina alterniflora*, *Phragmites australis*, and *Scirpus mariqueter* decreased with distance from the highest levels of suspended sediment concentrations (found at the marsh edge bordering tidal creeks or the mudflats); decreased with those species at the highest elevations, which experienced the lowest frequency and depth of tidal inundations; and increased with increasing plant biomass. *Spartina alterniflora*, which had the most sediment adhering to it, may contribute >10% of the total marsh surface sediment accretion by this process.

Salt marsh species also facilitate sediment accretion by decreasing current velocities and encouraging sediment to settle out of suspension. Current velocities can be reduced as the stems of tall marsh species induce hydraulic drag, with the effect of minimising re-suspension of sediment and encouraging deposition. Measured concentrations of suspended sediment in the water column have been shown to decrease from the open water or tidal creeks adjacent to the marsh edge, to the marsh interior, probably as a result of direct settling to the marsh surface by the influence of the marsh canopy.

Inundation and sediment deposition on the marsh surface is also assisted by tidal creeks which are a common feature of salt marshes. Their typically dendritic and meandering forms provide avenues for the tide to rise and flood the marsh surface, as well as to drain water, and they may facilitate higher amounts of sediment deposition than salt marsh bordering open ocean. Salt marshes do not however require tidal creeks to facilitate sediment flux over their surface although salt marshes with this morphology seem to be rarely studied.

The elevation of marsh species is important; those species at lower elevations experience longer and more frequent tidal floods and therefore have the opportunity for more sediment deposition to occur. Species at higher elevations can benefit from a greater chance of inundation at the highest tides when increased water depths and marsh surface flows can penetrate into the marsh interior.

Human Impacts

Spartina alterniflora (Saltmarsh Cordgrass). Native to the eastern seaboard of the United States. Considered a noxious weed in the Pacific Northwest

The coast is a highly attractive natural feature to humans through its beauty, resources, and accessibility. As of 2002, over half of the world's population was estimated to being living within 60 km of the coastal shoreline, making coastlines highly vulnerable to human impacts from daily activities that put pressure on these surrounding natural environments. In the past, salt marshes were perceived as coastal 'wastelands,' causing considerable loss and change of these ecosystems through land reclamation for agriculture, urban development, salt production and recreation. The indirect effects of human activities such as nitrogen loading also play a major role in the salt marsh area. Salt marshes can suffer from dieback in the high marsh and die-off in the low marsh.

Land Reclamation

Reclamation of land for agriculture by converting marshland to upland was historically a common practice. Dikes were often built to allow for this shift in land change and to provide flood protection further inland. In recent times intertidal flats have also been reclaimed. For centuries, livestock such as sheep and cattle grazed on the highly fertile salt marsh land. Land reclamation for agriculture has resulted in many changes such as shifts in vegetation structure, sedimentation, salinity, water flow, biodiversity loss and high nutrient inputs. There have been many attempts made to eradicate these problems for example, in New Zealand, the cordgrass *Spartina anglica* was introduced from England into the Manawatu River mouth in 1913 to try and reclaim the estuary land for farming. A shift in structure from bare tidal flat to pastureland resulted from increased sedimentation and the cordgrass extended out into other estuaries around New Zealand. Native plants and animals struggled to survive as non-natives out competed them. Efforts are now being made to remove these cordgrass species, as the damages are slowly being recognised.

In the Blyth estuary in Suffolk in eastern England, the mid-estuary reclamations (Angel and Bulcamp marshes) that were abandoned in the 1940s have been replaced by tidal flats with compacted soils from agricultural use overlain with a thin veneer of mud. Little vegetation colonisation has occurred in the last 60–75 years and has been attributed to a combination of surface elevations too low for pioneer species to develop, and poor drainage from the compacted agricultural soils acting as an aquaclude. Terrestrial soils of this nature need to adjust from fresh to saline interstitial water by a change in the chemistry and the structure of the soil, accompanied with fresh deposition of estuarine sediment, before salt marsh vegetation can establish. The vegetation structure, species richness, and plant community composition of salt marshes naturally regenerated on reclaimed agricultural land can be compared to adjacent reference salt marshes to assess the success of marsh regeneration.

Upstream Agriculture

Cultivation of land upstream from the salt marsh can introduce increased silt inputs and raise the rate of primary sediment accretion on the tidal flats, so that pioneer species can spread further onto the flats and grow rapidly upwards out of the level of tidal inundation. As a result, marsh surfaces in this regime may have an extensive cliff at their seaward edge. At the Plum Island estuary, Massachusetts (U.S.A), stratigraphic cores revealed that during the 18th and 19th century the marsh prograded over subtidal and mudflat environments to increase in area from 6 km² to 9 km² after European settlers deforested the land uptream and increased the rate of sediment supply.

Urban Development and Nitrogen Loading

Chaetomorpha linum is a common marine algae found in the salt marsh.

The conversion of marshland to upland for agriculture has in the past century been overshadowed by conversion for urban development. Coastal cities worldwide have encroached onto former salt marshes and in the U.S. the growth of cities looked to salt marshes for waste disposal sites. Estuarine pollution from organic, inorganic, and toxic substances from urban development or industrialisation is a worldwide problem and the sediment in salt marshes may entrain this pollution with toxic effects on floral and faunal species. Urban development of salt marshes has slowed since about 1970 owing to growing awareness by environmental groups that they provide beneficial ecosystem services. They are highly productive ecosystems, and when net productivity is measured in $g \, m^{-2} \, yr^{-1}$ they are equalled only by tropical rainforests. Additionally, they can help reduce wave erosion on sea walls designed to protect low-lying areas of land from wave erosion.

De-naturalisation of the landward boundaries of salt marshes from urban or industrial enchroachment can have negative effects. In the Avon-Heathcote estuary/Ihutai, New Zealand, species abundance and the physical properties of the surrounding margins were strongly linked, and the majority of salt marsh was found to be living along areas with natural margins in the Avon and Heathcote river outlets; conversely, artificial margins contained little marsh vegetation and restricted landward retreat. The remaining marshes surrounding these urban areas are also under immense pressure from the human population as human-induced nitrogen enrichment enters these habitats. Nitrogen loading through human-use indirectly affects salt marshes causing shifts in vegetation structure and the invasion of non-native species.

Human impacts such as sewage, urban run-off, agricultural and industrial wastes are running into the marshes from nearby sources. Salt marshes are nitrogen limited and with an increasing level of nutrients entering the system from anthropogenic effects, the plant species associated with salt marshes are being restructured through change in competition. For example, the New England salt marsh is experiencing a shift in vegetation structure where *S. alterniflora* is spreading from the lower marsh where it predominately resides up into the upper marsh zone. Additionally, in the same marshes, the reed *Phragmites australis* has been invading the area expanding to lower marshes and becoming a dominant species. *P. australis* is an aggressive halophyte that can invade disturbed areas in large numbers outcompeting native plants. This loss in biodiversity is not only seen in flora assemblages but also in many animals such as insects and birds as their habitat and food resources are altered.

Sea Level Rise

Due to the melting of Arctic sea ice, as a result of global warming, sea levels have begun to rise. As with all coastlines, this rise in water levels are predicted to negatively effect salt marshes, by flooding and eroding them.

Mosquito Control

Earlier in the 20th century, it was believed that draining salt marshes would help reduce mosquito populations. In many locations, particularly in the northeastern United

States, residents and local and state agencies dug straight-lined ditches deep into the marsh flats. The end result, however, was a depletion of killifish habitat. The killifish is a mosquito predator, so the loss of habitat actually led to higher mosquito populations, and adversely affected wading birds that preyed on the killifish. These ditches can still be seen, despite some efforts to refill the ditches.

Crab Herbivory and Bioturbation

The tunnelling mud crab *Helice crassa* of New Zealand fills a special niche in the salt marsh ecosystem.

Increased nitrogen uptake by marsh species into their leaves can prompt greater rates of length-specific leaf growth, and increase the herbivory rates of crabs. The burrowing crab *Neohelice granulata* frequents SW Atlantic salt marshes where high density populations can be found among populations of the marsh species *Spartina densiflora* and *Sarcocornia perennis*. In Mar Chiquita lagoon, north of Mar del Plata, Argentina, *Neohelice granulata* herbivory increased as a likely response to the increased nutrient value of the leaves of fertilised *Spartina densiflora* plots, compared to non-fertilised plots. Regardless of whether the plots were fertilised or not, grazing by *Neohelice granulata* also reduced the length specific leaf growth rates of the leaves in summer, while increasing their length-specific senescence rates. This may have been assisted by the increased fungal effectiveness on the wounds left by the crabs.

The salt marshes of Cape Cod, Massachusetts (U.S.A), are experiencing creek bank die-offs of *Spartina* spp. (cordgrass) that has been attributed to herbivory by the crab *Sesarma reticulatum*. At 12 surveyed Cape Cod salt marsh sites, 10% - 90% of creek banks experienced die-off of cordgrass in association with a highly denuded substrate and high density of crab burrows. Populations of *Sesarma reticulatum* are increasing, possibly as a result of the degradation of the coastal food web in the region. The bare areas left by the intense grazing of cordgrass by *Sesarma reticulatum* at Cape Cod are suitable for occupation by another burrowing crab, *Uca pugnax*, which are not known to consume live macrophytes. The intense bioturbation of salt marsh sediments from this crab's burrowing activity has been shown to dramatically reduce the success of *Spartina alterniflora* and *Suaeda maritima* seed germination and established seedling survival, either by burial or exposure of seeds, or uprooting or burial of established seedlings. However, bioturbation by crabs may also have a positive effect. In New Zealand, the tunnelling mud crab *Helice crassa* has been given the stately name of an 'ecosystem engineer' for its ability to construct new habitats and alter the access of nutrients to other species. Their burrows provide an avenue for the transport of dissolved oxygen in the burrow water through the oxic sediment of the burrow walls and into the surrounding anoxic sediment, which creates the perfect habitat for special nitrogen cycling bacteria. These nitrate reducing (denitrifying) bacteria quickly consume the dissolved oxygen entering into the burrow walls to create the oxic mud layer that is thinner than that at the mud surface. This allows a more direct diffusion path for the export of nitrogen (in the form of gaseous nitrogen (N_2)) into the flushing tidal water.

Restoration and Management

Glasswort (*Salicornia spp.*) a species endemic to the high marsh zone.

The perception of bay salt marshes as a coastal 'wasteland' has since changed, acknowledging that they are one of the most biologically productive habitats on earth, rivalling tropical rainforests. Salt marshes are ecologically important providing habitats for native migratory fish and acting as sheltered feeding and nursery grounds. They are now protected by legislation in many countries to look after these ecologically important habitats. In the United States and Europe, they are now accorded to a high level of protection by the Clean Water Act and the Habitats Directive respectively. With the impacts of this habitat and its importance now realised, a growing interest in restoring salt marshes, through managed retreat or the reclamation of land has been established. However, many Asian countries such as China are still to recognise the value of marshlands. With their ever-growing populations and intense development along the coast, the value of salt marshes tends to be ignored and the land continues to be reclaimed.

Bakker et al. (1997) suggests two options available for restoring salt marshes. The first is to abandon all human interference and leave the salt marsh to complete its natural development. These types of restoration projects are often unsuccessful as vegetation tends to struggle to revert to its original structure and the natural tidal cycles are shifted due to land changes. The second option suggested by Bakker et al. (1997) is to restore the destroyed habitat into its natural state either at the original site or as a replacement at a different site. Under natural conditions, recovery can take 2–10 years or even longer depending on the nature and degree of the disturbance and the relative maturity of the marsh involved. Marshes in their pioneer stages of development will recover more rapidly than mature marshes as they are often first to colonize the land. It is important to note, that restoration can often be sped up through the replanting of native vegetation.

Common reed (*Phragmites australis*) an invasive species in degraded marshes in the northeastern United States.

This last approach is often the most practiced and generally more successful than allowing the area to naturally recover on its own. The salt marshes in the state of Connecticut in the United States have long been an area lost to fill and dredging. As of 1969, the Tidal Wetland Act was introduced that seized this practice, but despite the introduction of the act, the system was still degrading due to alterations in tidal flow. One area in Connecticut is the marshes on Barn Island. These marshes were diked then impounded with salt and brackish marsh during 1946-1966. As a result, the marsh shifted to a freshwater state and became dominated by the invasive species *P. australis*, *Typha angustifolia* and *T. latifolia* that have little ecological connection to the area.

By 1980, a restoration programme was put in place that has now been running for over 20 years. This programme has aimed to reconnect the marshes by returning tidal flow along with the ecological functions and characteristics of the marshes back to their original state. In the case of Barn Island, declines in the invasive species have initiated, re-establishing the tidal-marsh vegetation along with animal species such as fish and insects. This example highlights that considerable time and effort is needed to effectively restore salt marsh systems. Times in marsh recovery can depend on the development stage of the marsh; type and extent of the disturbance; geographical location; and the environmental and physiological stress factors to the marsh-associated flora and fauna.

Although much effort has gone into restoring salt marshes worldwide, further research is needed. There are many setbacks and problems associated with marsh restoration that requires careful long-term monitoring. Information on all components of the salt marsh ecosystem should be understood and monitored from sedimentation, nutrient, and tidal influences, to behaviour patterns and tolerances of both flora and fauna species. Once we have a better understanding of these processes and not just locally, but over a global scale, we can then suggest more sound and practical management and restoration efforts that can be used to preserve our valuable marshes and put them back to their original state.

While humans are situated along coastlines, there will always be the possibility of human-induced disturbances despite the number of restoration efforts we plan to implement. Dredging, pipelines for offshore petroleum resources, highway construction, accidental toxic spills or just plain carelessness are examples that will for some time now and into the future be the major influences of salt marsh degradation.

Atlantic ribbed mussell, found in the low marsh

In addition to restoring and managing salt marsh systems based on scientific principles, the opportunity should be taken to educate public audiences of their importance biologically and their purpose as serving as a natural buffer for flood protection. Because salt marshes are often located next to urban areas, they are likely to receive more visitors than remote wetlands. By physically seeing the marsh, people are more likely to take notice and be more aware of the environment around them. An example of public involvement occurred at the Famosa Slough State Marine Conservation Area in San Diego, where a "friends" group worked for over a decade in trying to prevent the area from being developed. Eventually, the 5 hectare site was bought by the City and the group worked together to restore the area. The project involved removing invasive species and replanting with natives, along with public talks to other locals, frequent bird walks and clean-up events.

Research Methods

There is a diverse range and combination of methodologies employed to understand the hydrological dynamics in salt marshes and their ability to trap and accrete sediment. Sediment traps are often used to measure rates of marsh surface accretion when short term deployments (e.g. less than one month) are required. These circular traps consist of pre-weighed filters that are anchored to the marsh surface, then dried in a laboratory and re-weighed to determine the total deposited sediment. For longer term studies (e.g. more than one year) researchers may prefer to measure sediment accretion with marker horizon plots. Marker horizons consist of a mineral such as feldspar that is buried at a known depth within wetland substrates to record the increase in overlying substrate over long time periods. In order to gauge the amount of sediment suspended in the water column, manual or automated samples of tidal water can be poured through pre-weighed filters in a laboratory then dried to determine the amount of sediment per volume of water. Another method for estimating suspended sediment concentrations is by measuring the turbidity of the water using optical backscatter probes, which can be calibrated against water samples containing a known suspended sediment concentration to establish a regression relationship between the two. Marsh surface elevations may be measured with a stadia rod and transit, electronic theodolite, Real-Time Kinematic Global Positioning System, laser level or electronic distance meter (total station). Hydrological dynamics include water depth, measured automatically with a pressure transducer, or with a marked wooden stake, and water velocity, often using electromagnetic current meters.

Estuary

An estuary is a partially enclosed coastal body of brackish water with one or more rivers or streams flowing into it, and with a free connection to the open sea.

Estuaries form a transition zone between river environments and maritime environments. They are subject both to marine influences—such as tides, waves, and the influx of saline water—and to riverine influences—such as flows of fresh water and sediment. The inflows of both sea water and fresh water provide high levels of nutrients both in the water column and in sediment, making estuaries among the most productive natural habitats in the world.

Most existing estuaries formed during the Holocene epoch with the flooding of river-eroded or glacially scoured valleys when the sea level began to rise about 10,000–12,000 years ago. Estuaries are typically classified according to their geomorphological features or to water-circulation patterns. They can have many different names, such as bays, harbors, lagoons, inlets, or sounds, although some of these water bodies do not strictly meet the above definition of an estuary and may be fully saline.

The banks of many estuaries are amongst the most heavily populated areas of the world, with about 60% of the world's population living along estuaries and the coast. As a result, many estuaries suffer degradation by many factors, including sedimentation from soil erosion from deforestation, overgrazing, and other poor farming practices; overfishing; drainage and filling of wetlands; eutrophication due to excessive nutrients from sewage and animal wastes; pollutants including heavy metals, polychlorinated biphenyls, radionuclides and hydrocarbons from sewage inputs; and diking or damming for flood control or water diversion.

Definition

Hudson River estuary waterways around New York City, United States: 1. Hudson River, 2. East River, 3. Long Island Sound, 4. Newark Bay, 5. Upper New York Bay, 6. Lower New York Bay, separated from Upper New York Bay by the Narrows strait, 7. Jamaica Bay, and 8. Atlantic Ocean.

River Exe estuary

River Nith estuary

Estuary mouth located in Darwin, Northern Territory, Australia

A crowded estuary mouth in Paravur near the city of Kollam, India

Estuary mouth

Río de la Plata estuary

Estuary mouth of the Yachats River in Yachats, Oregon

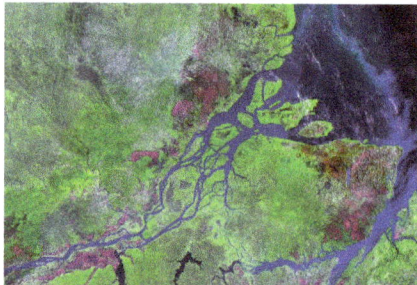
Amazon estuary

The word "estuary" is derived from the Latin word *aestuarium* meaning tidal inlet of the sea, which in itself is derived from the term *aestus*, meaning tide. There have been many definitions proposed to describe an estuary. The most widely accepted definition is: "a semi-enclosed coastal body of water, which has a free connection with the open sea, and within sea water is measurably diluted with freshwater derived from land drainage". However, this definition excludes a number of coastal water bodies such as coastal lagoons and brackish seas. A more comprehensive definition of an estuary is "a semi-enclosed body of water connected to the sea as far as the tidal limit or the salt intrusion limit and receiving freshwater runoff; however the freshwater inflow may not be perennial, the connection to the sea may be closed for part of the year and tidal influence may be negligible". This broad definition also includes fjords, lagoons, river mouths, and tidal creeks. An estuary is a dynamic ecosystem with a connection with the open sea through which the sea water enters with the rhythm of the tides. The sea water entering the estuary is diluted by the fresh water flowing from rivers and streams. The pattern of dilution varies between different estuaries and depends on the volume of fresh water, the tidal range, and the extent of evaporation of the water in the estuary.

Classification Based on Geomorphology

Drowned River Valleys

Drowned river valleys also known as coastal plain estuaries. In place where the sea level is rising relative to the land, sea water progressively penetrates into river valleys and the topography of the estuary remains similar to that of a river valley. this is the most common type of estuary in temperate climates. Well-studied estuaries include the Severn Estuary in the United Kingdom and the Ems Dollard along the Dutch-German border.

The width-to-depth ratio of these estuaries is typically large, appearing wedge-shaped (in cross-section) in the inner part and broadening and deepening seaward. Water depths rarely exceed 30 m (100 ft). Examples of this type of estuary in the U.S. are the Hudson River, Chesapeake Bay, and Delaware Bay along the Mid-Atlantic coast, and Galveston Bay and Tampa Bay along the Gulf Coast. San Francisco Bay is another good example of a drowned river valley.

Lagoon-type or Bar-built

Bar-built estuaries are found in place where the deposition of sediment has kept pace with rising sea level so that the estuaries are shallow and separated from the sea by sand spits or barrier islands. They are partially common in tropical and subtropical locations.

These estuaries are semi-isolated from ocean waters by barrier beaches (barrier islands and barrier spits). Formation of barrier beaches partially encloses the estuary, with only narrow inlets allowing contact with the ocean waters. Bar-built estuaries typically develop on gently sloping plains located along tectonically stable edges of continents and marginal sea coasts. They are extensive along the Atlantic and Gulf coasts of the U.S. in areas with active coastal deposition of sediments and where tidal ranges are less than 4 m (13 ft). The barrier beaches that enclose bar-built estuaries have been developed in several ways:

- building up of offshore bars by wave action, in which sand from the sea floor is deposited in elongated bars parallel to the shoreline,

- reworking of sediment discharge from rivers by wave, current, and wind action into beaches, overwash flats, and dunes,

- engulfment of mainland beach ridges (ridges developed from the erosion of coastal plain sediments around 5000 years ago) due to sea level rise and resulting in the breaching of the ridges and flooding of the coastal lowlands, forming shallow lagoons, and

- elongation of barrier spits from the erosion of headlands due to the action of longshore currents, with the spits growing in the direction of the littoral drift.

Barrier beaches form in shallow water and are generally parallel to the shoreline, resulting in long, narrow estuaries. The average water depth is usually less than 5 m (16 ft), and rarely exceeds 10 m (33 ft). Examples of bar-built estuaries are Barnegat Bay, New Jersey; Laguna Madre, Texas; and Pamlico Sound, North Carolina.

Fjord-type

fjords are formed where pleistocene glaciers deepened and widened existing river valleys so that the become U-shaped in cross sections. At their mouths there are typically rocks bars or sills of glacial deposits, which have the effects of modifying the estuarine circulation.

Fjord-type estuaries are formed in deeply eroded valleys formed by glaciers. These U-shaped estuaries typically have steep sides, rock bottoms, and underwater sills contoured by glacial movement. The estuary is shallowest at its mouth, where terminal glacial moraines or rock bars form sills that restrict water flow. In the upper reaches of the estuary, the depth can exceed 300 m (1,000 ft). The width-to-depth ratio is generally small. In estuaries with very shallow sills, tidal oscillations only affect the water down to the depth of the sill, and the waters deeper than that may remain stagnant for a very long time, so there is only an occasional exchange of the deep water of the estuary with the ocean. If the sill depth is deep, water circulation is less restricted, and there is a slow but steady exchange of water between the estuary and the ocean. Fjord-type estuaries can be found along the coasts of Alaska, the Puget Sound region of western

Washington state, British Columbia, eastern Canada, Greenland, Iceland, New Zealand, and Norway.

Tectonically Produced

These estuaries are formed by subsidence or land cut off from the ocean by land movement associated with faulting, volcanoes, and landslides. Inundation from eustatic sea level rise during the Holocene Epoch has also contributed to the formation of these estuaries. There are only a small number of tectonically produced estuaries; one example is the San Francisco Bay, which was formed by the crustal movements of the San Andreas fault system causing the inundation of the lower reaches of the Sacramento and San Joaquin rivers.

Classification Based on Water Circulation

Salt Wedge

In this type of estuary, river output greatly exceeds marine input and tidal effects have a minor importance. Fresh water floats on top of the seawater in a layer that gradually thins as it moves seaward. The denser seawater moves landward along the bottom of the estuary, forming a wedge-shaped layer that is thinner as it approaches land. As a velocity difference develops between the two layers, shear forces generate internal waves at the interface, mixing the seawater upward with the freshwater. An example of a salt wedge estuary is the Mississippi River.

Partially Mixed

As tidal forcing increases, river output becomes less than the marine input. Here, current induced turbulence causes mixing of the whole water column such that salinity varies more longitudinally rather than vertically, leading to a moderately stratified condition. Examples include the Chesapeake Bay and Narragansett Bay.

Well-Mixed

Tidal mixing forces exceed river output, resulting in a well mixed water column and the disappearance of the vertical salinity gradient. The freshwater-seawater boundary is eliminated due to the intense turbulent mixing and eddy effects. The lower reaches of Delaware Bay and the Raritan River in New Jersey are examples of vertically homogenous estuaries.

Inverse

Inverse estuaries occur in dry climates where evaporation greatly exceeds the inflow of fresh water. A salinity maximum zone is formed, and both riverine and oceanic water flow close to the surface towards this zone. This water is pushed downward and spreads along the bottom in both the seaward and landward direction. An example of an inverse estuary is Spencer Gulf, South Australia.

Intermittent

Estuary type varies dramatically depending on freshwater input, and is capable of changing from a wholly marine embayment to any of the other estuary types.

Physiochemical Variation

The most important variable characteristics of estuary water are the concentration of dissolved oxygen, salinity and sediment load. There is extreme spatial variability in salinity, with a range of near zero at the tidal limit of tributary rivers to 3.4% at the estuary mouth. At any one point the salinity will vary considerably over time and seasons, making it a harsh environment for organisms. Sediment often settles in intertidal mudflats which are extremely difficult to colonize. No points of attachment exist for algae, so vegetation based habitat is not established. Sediment can also clog feeding and respiratory structures of species, and special adaptations exist within mudflat species to cope with this problem. Lastly, dissolved oxygen variation can cause problems for life forms. Nutrient-rich sediment from man-made sources can promote primary production life cycles, perhaps leading to eventual decay removing the dissolved oxygen from the water; thus hypoxic or anoxic zones can develop.

Implications for Marine Life

Estuaries provide habitats for a large number of organisms and support very high productivity. Estuaries provide habitats for many fish nurseries, depending upon their locations in the world, such as salmon and sea trout. Also, migratory bird populations, such as the black-tailed godwit, make essential use of estuaries.

Two of the main challenges of estuarine life are the variability in salinity and sedimentation. Many species of fish and invertebrates have various methods to control or conform to the shifts in salt concentrations and are termed osmoconformers and osmoregulators. Many animals also burrow to avoid predation and to live in the more stable sedimental environment. However, large numbers of bacteria are found within the sediment which have a very high oxygen demand. This reduces the levels of oxygen within the sediment often resulting in partially anoxic conditions, which can be further exacerbated by limited water flux.

Phytoplankton are key primary producers in estuaries. They move with the water bodies and can be flushed in and out with the tides. Their productivity is largely dependent upon the turbidity of the water. The main phytoplankton present are diatoms and dinoflagellates which are abundant in the sediment.

It is important to remember that a primary source of food for many organisms on estuaries, including bacteria, is detritus from the settlement of the sedimentation.

Human Impact

Of the thirty-two largest cities in the world, twenty-two are located on estuaries. For example, New York City is located at the mouth of the Hudson River estuary.

As ecosystems, estuaries are under threat from human activities such as pollution and overfishing. They are also threatened by sewage, coastal settlement, land clearance and much more. Estuaries are affected by events far upstream, and concentrate materials such as pollutants and sediments. Land run-off and industrial, agricultural, and domestic waste enter rivers and are discharged into estuaries. Contaminants can be introduced which do not disintegrate rapidly in the marine environment, such as plastics, pesticides, furans, dioxins, phenols and heavy metals.

Such toxins can accumulate in the tissues of many species of aquatic life in a process called bioaccumulation. They also accumulate in benthic environments, such as estuaries and bay muds: a geological record of human activities of the last century.

For example, Chinese and Russian industrial pollution, such as phenols and heavy metals, has devastated fish stocks in the Amur River and damaged its estuary soil.

Estuaries tend to be naturally eutrophic because land runoff discharges nutrients into estuaries. With human activities, land run-off also now includes the many chemicals used as fertilizers in agriculture as well as waste from livestock and humans. Excess oxygen-depleting chemicals in the water can lead to hypoxia and the creation of dead zones. This can result in reductions in water quality, fish, and other animal populations.

Overfishing also occurs. Chesapeake Bay once had a flourishing oyster population that has been almost wiped out by overfishing. Oysters filter these pollutants, and either eat them or shape them into small packets that are deposited on the bottom where they are harmless. Historically the oysters filtered the estuary's entire water volume of excess nutrients every three or four days. Today that process takes almost a year, and sediment, nutrients, and algae can cause problems in local waters.

Notable Examples

- Amazon River
- Chesapeake Bay
- Delaware Bay
- Drake's Estero

- Gippsland Lakes
- Gironde
- Golden Horn
- Great Bay
- Gulf of Saint Lawrence
- Hampton Roads
- Humber
- Laguna Madre
- Lake Borgne
- Lake Merritt
- Lake Pontchartrain
- Long Island Sound
- Mobile Bay
- Narragansett Bay
- New York-New Jersey Harbor
- Ob River
- Puget Sound
- Pamlico Sound
- Port Jackson (Sydney Harbour)
- Rio de la Plata
- San Francisco Bay
- Severn Estuary
- Shannon Estuary
- Spencer Gulf
- Tampa Bay
- Thames Estuary
- Western Scheldt

Intertidal Zone

The intertidal zone, also known as the foreshore and seashore and sometimes referred to as the littoral zone, is the area that is above water at low tide and under water at high tide (in other words, the area between tide marks). This area can include many different types of habitats, with many types of animals, such as starfish, sea urchins, and numerous species of coral. The well-known area also includes steep rocky cliffs, sandy beaches, or wetlands (e.g., vast mudflats). The area can be a narrow strip, as in Pacific islands that have only a narrow tidal range, or can include many meters of shoreline where shallow beach slopes interact with high tidal excursion.

Organisms in the intertidal zone are adapted to an environment of harsh extremes. The intertidal zone is also home to many several species from different taxa including Porifera, Annelids, Coelenterates, Mollusks, Crustaceans, Arthropods, etc. Water is available regularly with the tides but varies from fresh with rain to highly saline and dry salt with drying between tidal inundations. The action of waves can dislodge residents from the littoral zone. With the intertidal zone's high exposure to the sun, the temperature range can be anything from very hot with full sun to near freezing in colder climates. Some microclimates in the littoral zone are ameliorated by local features and larger plants such as mangroves. Adaptation in the littoral zone allows the use of nutrients supplied in high volume on a regular basis from the sea, which is actively moved to the zone by tides. Edges of habitats, in this case land and sea, are themselves often significant ecologies, and the littoral zone is a prime example of a sea and more things

A typical rocky shore can be divided into a spray zone or splash zone (also known as the supratidal zone), which is above the spring high-tide line and is covered by water only during storms, and an intertidal zone, which lies between the high and low tidal extremes. Along most shores, the intertidal zone can be clearly separated into the following subzones: high tide zone, middle tide zone, and low tide zone. The intertidal zone is one of a number of marine biomes or habitats, including estuaries, neritic, surface and deep zones.

Zonation

Marine biologists divide the intertidal region into three zones (low, middle, and high), based on the overall average exposure of the zone. The low intertidal zone, which borders on the shallow subtidal zone, is only exposed to air at the lowest of low tides and is primarily marine in character. The mid intertidal zone is regularly exposed and submerged by average tides. The high intertidal zone is only covered by the highest of the high tides, and spends much of its time as terrestrial habitat. The high intertidal zone borders on the splash zone (the region above the highest still-tide level, but which receives wave splash). On shores exposed to heavy wave action, the intertidal zone will be influenced by waves, as the spray from breaking waves will extend the intertidal zone.

Tide pools at Pillar Point showing zonation on the edge of the rock ledge

A rock, seen at low tide, exhibiting typical intertidal zonation, Kalaloch, Washington, western USA.

Depending on the substratum and topography of the shore, additional features may be noticed. On rocky shores, tide pools form in depressions that fill with water as the tide rises. Under certain conditions, such as those at Morecambe Bay, quicksand may form.

Low Tide Zone (Lower Littoral)

This subregion is mostly submerged - it is only exposed at the point of low tide and for a longer period of time during extremely low tides. This area is teeming with life; the most notable difference with this subregion to the other three is that there is much more marine vegetation, especially seaweeds. There is also a great biodiversity. Organisms in this zone generally are not well adapted to periods of dryness and temperature extremes. Some of the organisms in this area are abalone, sea anemones, brown seaweed, chitons, crabs, green algae, hydroids, isopods, limpets, mussels, nudibranchs, sculpin, sea cucumber, sea lettuce, sea palms, starfish, sea urchins, shrimp, snails, sponges, surf grass, tube worms, and whelks. Creatures in this area can grow to larger sizes because there is more available energy in the localized ecosystem. Also, marine vegetation can grow to much greater sizes than in the other three intertidal subregions due to the better water coverage. The water is shallow enough to allow plenty of light to reach the vegetation to allow substantial photosynthetic activity, and the salinity is at almost normal levels. This area is also protected from large predators such as fish because of the wave action and the relatively shallow water.

Ecology

A California tide pool in the low tide zone

The intertidal region is an important model system for the study of ecology, especially on wave-swept rocky shores. The region contains a high diversity of species, and the zonation created by the tides causes species ranges to be compressed into very narrow bands. This makes it relatively simple to study species across their entire cross-shore range, something that can be extremely difficult in, for instance, terrestrial habitats that can stretch thousands of kilometres. Communities on wave-swept shores also have high turnover due to disturbance, so it is possible to watch ecological succession over years rather than decades.

Since the foreshore is alternately covered by the sea and exposed to the air, organisms living in this environment must have adaptions for both wet and dry conditions. Hazards include being smashed or carried away by rough waves, exposure to dangerously high temperatures, and desiccation. Typical inhabitants of the intertidal rocky shore include urchins, sea anemones, barnacles, chitons, crabs, isopods, mussels, starfish, and many marine gastropod molluscs such as limpets and whelks.

Legal Issues

As with the dry sand part of a beach, legal and political disputes can arise over the ownership and use of the foreshore. One recent example is the New Zealand foreshore and seabed controversy. In legal discussions the foreshore is often referred to as the *wet-sand area.*

For privately owned beaches in the United States, some states such as Massachusetts use the low water mark as the dividing line between the property of the State and that of the beach owner. Other states such as California use the high-water mark.

In the UK, the foreshore is generally deemed to be owned by the Crown although there are notable exceptions, especially what are termed *several fisheries,* which can be historic deeds to title, dating back to King John's time or earlier, and the Udal Law, which applies generally in Orkney and Shetland.

In Greece, according to the L. 2971/01, the foreshore zone is defined as the area of the coast that might be reached by the maximum climbing of the waves on the coast (maximum wave run-up on the coast) in their maximum capacity (maximum referring to the "usually maximum winter waves" and of course not to exceptional cases, such as tsunamis etc.). The foreshore zone, apart of the exceptions of the law, is public, and permanent constructions are not allowed on it.

Other Media

The Intertidal Zone was used as a title for Stephen Hillenburg's old comic strip. The comic strip starred "Bob The Sponge", who would later go on to become SpongeBob SquarePants.

Lagoon

A lagoon is a shallow body of water separated from a larger body of water by barrier islands or reefs. Lagoons are commonly divided into coastal lagoons and atoll lagoons. They have also been identified as occurring on mixed-sand and gravel coastlines. There is an overlap between bodies of water classified as coastal lagoons and bodies of water classified as estuaries. Lagoons are common coastal features around many parts of the world.

Lagoons can also be man-made and used for wastewater treatment, as is the case for e.g. aerated lagoons and anaerobic lagoons.

Definition

Lagoons are shallow, often elongated bodies of water separated from a larger body of water by a shallow or exposed shoal, coral reef, or similar feature. Some authorities include fresh water bodies in the definition of "lagoon", while others explicitly restrict "lagoon" to bodies of water with some degree of salinity. The distinction between "lagoon" and "estuary" also varies between authorities. Richard A. Davis Jr. restricts "lagoon" to bodies of water with little or no fresh water inflow, and little or no tidal flow, and calls any bay that receives a regular flow of fresh water an "estuary". Davis does state that the terms "lagoon" and "estuary" are "often loosely applied, even in scientific literature." Timothy M. Kusky characterizes lagoons as normally being elongated parallel to the coast, while estuaries are usually drowned river valleys, elongated perpendicular to the coast. When used within the context of a distinctive portion of coral reef ecosystems, the term "lagoon" is synonymous with the term "back reef" or "backreef", which is more commonly used by coral reef scientists to refer to the same area. Coastal lagoons are classified as inland bodies of water.

Garabogaz-Göl lagoon in Turkmenistan

Many lagoons do not include "lagoon" in their common names. Albemarle and Pamlico sounds in North Carolina, Great South Bay between Long Island and the barrier beaches of Fire Island in New York, Isle of Wight Bay, which separates Ocean City, Maryland from the rest of Worcester County, Maryland, Banana River in Florida, Lake Illawarra in New South Wales, Montrose Basin in Scotland, and Broad Water in Wales have all been classified as lagoons, despite their names. In England, The Fleet at Chesil Beach has also been described as a lagoon.

In Latin America, the term "laguna", which lagoon translates to, is often used to describe a lake, such as Laguna Catemaco. In Portuguese, "lagoa" may be a body of shallow sea water, but also a relatively small freshwater lake not linked to the sea.

Etymology

Lagoon is derived from the Italian *laguna*, which refers to the waters around Venice, the Lagoon of Venice. *Laguna* is attested in English by at least 1612, and had been Anglicized to "lagune" by 1673. In 1697 William Dampier referred to a "Lagune or Lake of Salt water" on the coast of Mexico. Captain James Cook described an island "of Oval form with a Lagoon in the middle" in 1769.

Atoll Lagoons

Atoll lagoons form as coral reefs grow upwards while the islands that the reefs surround subside, until eventually only the reefs remain above sea level. Unlike the lagoons that form shoreward of fringing reefs, atoll lagoons often contain some deep (>20m) portions.

Coastal Lagoons

Coastal lagoon landscapes around the island of Hiddensee near Stralsund, Germany. Many similar coastal lagoons can be found around the Western Pomerania Lagoon Area National Park.

Coastal lagoons form along gently sloping coasts where barrier islands or reefs can develop off-shore, and the sea-level is rising relative to the land along the shore (either because of an intrinsic rise in sea-level, or subsidence of the land along the coast). Coastal lagoons do not form along steep or rocky coasts, or if the range of tides is more than 4 metres (13 ft). Due to the gentle slope of the coast, coastal lagoons are shallow. They are sensitive to changes in sea level due to global warming. A relative drop in sea level may leave a lagoon largely dry, while a rise in sea level may let the sea breach or destroy barrier islands, and leave reefs too deep under water to protect the lagoon. Coastal lagoons are young and dynamic, and may be short-lived in geological terms. Coastal lagoons are common, occurring along nearly 15 percent of the world's shorelines. In the United States, lagoons are found along more than 75 percent of the eastern and Gulf coasts.

Coastal lagoons are usually connected to the open ocean by inlets between barrier islands. The number and size of the inlets, precipitation, evaporation, and inflow of fresh water all affect the nature of the lagoon. Lagoons with little or no interchange with the open ocean, little or no inflow of fresh water, and high evaporation rates, such as Lake St. Lucia, in South Africa, may become highly saline. Lagoons with no connection to the open ocean and significant inflow of fresh water, such as the Lake Worth Lagoon in Florida in the middle of the 19th century, may be entirely fresh. On the other hand, lagoons with many wide inlets, such as the Wadden Sea, have strong tidal currents and mixing. Coastal lagoons tend to accumulate sediments from inflowing rivers, from run-off from the shores of the lagoon, and from sediment carried into the lagoon through inlets by the tide. Large quantities of sediment may be occasionally be deposited in a lagoon when storm waves overwash barrier islands. Mangroves and marsh plants can facilitate the accumulation of sediment in a lagoon. Benthic organisms may stabilize or destabilize sediments.

River-mouth Lagoons on Mixed Sand and Gravel Beaches

River-mouth lagoons on mixed sand and gravel (MSG) beaches form at the river-coast interface where a typically braided, although sometimes meandering, river interacts with a coastal environment that is significantly affected by longshore drift. The lagoons which form on the MSG coastlines are common on the east coast of the South Island of New Zealand and have long been referred to as hapua by the Māori. This classification differentiates hapua from similar lagoons located on the New Zealand coast termed waituna. Hapua are often located on paraglacial coastal areas where there is a low level of coastal development and minimal population density. Hapua form as the river carves out an elongated coast-parallel area, blocked from the sea by a MSG barrier which constantly alters its shape and volume due to longshore drift. Longshore drift continually extends the barrier behind which the hapua forms by transporting sediment along the coast. Hapua are defined as a narrow shore-parallel extensions of the coastal riverbed. They discharge the majority of stored water to the ocean via an ephemeral and highly mobile drainage channel or outlet. The remainder percolates through the MSG barrier due to its high levels of permeability. Hapua systems are driven by a wide range of dynamic processes that are generally classified as fluvial or marine; changes in the balance between these processes as well as the antecedent barrier conditions can cause shifts in the morphology of the hapua, in particular the barrier. New Zealand examples include the Rakaia, Ashburton and Hurunui river-mouths.

Hapua Environment

Hapua have been identified as establishing in the Canterbury Bight coastal region on the east coast of the South Island. They are often found in areas of coarse-grained sediment where contributing rivers have moderately steep bed gradients. MSG beaches in the Canterbury Bight region contain a wide range of sediment sizes from sand to boulders and are exposed to the high energy waves that make up an east coast swell environment. MSG beaches are reflective rather than dissipative energy zones due to their morphological characteristics. They have a steep foreshore which is known as the 'engine room' of the beach profile. In this zone, swash and backwash are dominating processes alongside longshore transport. MSG beaches do not have a surf zone; instead a single line of breakers is visible in all sea conditions. Hapua are associated with MSG beaches as the variation in sediment size allows for the barrier to be permeable.

The east coast of the South Island has been identified as being in a period of chronic erosion of approximately 0.5 metres per year. This erosion trend is a result of a number of factors. According to the classification scheme of Zenkovich, the rivers on the east coast can be described as 'small'; this classification is not related to their flow rate but to the insufficient amount of sediment that they transport to the coast to nourish it. The sediment provided is not adequate to nourish the coast against its typical high energy waves and strong longshore drift. These two processes constantly remove sediment depositing it either offshore or further up drift. As the coastline becomes eroded the hapua have been 'rolling back' by eroding the backshore to move landwards.

Hapua or river-mouth lagoons form in micro-tidal environments. A micro-tidal environment is where the tidal range (distance between low tide and high tide) is less than two metres. Tidal currents in a micro-tidal zone are less than those found on meso-tidal (two – four metres) and macro-tidal (greater than four metres) coastlines. Hapua form in this type of tidal environment as the tidal currents are unable to compete with the powerful freshwater flows of the rivers therefore there is no negligible tidal penetration to the lagoon. A fourth element of the environment in which hapua form is the strong longshore drift component. Longshore or littoral drift is the transportation of sediments along the coast at an angle to the shoreline. In the Canterbury Bight coastal area; the dominant swell direction is northwards from the Southern Ocean. Therefore, the principal movement of sediment via longshore drift is north towards Banks Peninsula. Hapua are located in areas dominated by longshore drift; because it aids the formation of the barrier behind which the hapua is sited.

A hapua also requires sediment to form the lagoon barrier. Sediment which nourishes the east coast of New Zealand can be sourced from three different areas. Material from the highly erodible Southern Alps is removed via weathering; then carried across the Canterbury Plains by various braided rivers to the east coast beaches. The second source of sediment is the high cliffs which are located in the hinterland of lagoons. These can be eroded during the occurrence of high river flow or sea storm events. Beaches further south provide nourishment to the northern coast via longshore transport.

Hapua Characteristics

Hapua have a number of characteristics which includes shifts between a variety of morphodynamic states due to changes in the balance between marine and fluvial processes as well as the antecedent barrier conditions. The MSG barrier constantly changes size and shape as a result of the longshore drift. Water stored in the hapua drains to the coast predominately though an outlet; although it can also seep through the barrier depending on the permeability of the material.

Changes in the level of the lagoon water do not occur as a result of saltwater or tidal intrusion. Water in a hapua is predominately freshwater originating from the associated river. Hapua are non-estuarine, there is no tidal inflow however the tide does have an effect on the level of water in the lagoon. As the tide reaches its peak, the lagoon water has a much smaller amount of barrier to permeate through so the lagoon level rises. This is related to a physics theory known as hydraulic head. The lagoon level has a similar sinusoidal wave shape as the tide but reaches its peak slightly later. In general, any saltwater intrusion into the hapua will only occur during a storm via wave overtopping or sea spray.

Hapua can act as both a source and sink of sediment. The majority of sediment in the hapua is fluvial sourced. During medium to low river flows, coarser sediment generally collects in the hapua; while some of the finer sediment can be transported through the outlet to the coast. During flood events the hapua is 'flushed out' with larger amounts of sediment transferred through the outlet. This sediment can be deposited offshore or downdrift of the hapua replenishing the undernourished beach. If a large amount of material is released to the coast at one time it can be identified as a 'slug'. These can often be visible from aerial photographs.

Antecedent barrier conditions combined with changes in the balance between marine and fluvial processes results in shifts between a variety of morphological states in a hapua or river-mouth lagoon on a MSG beach. Marine processes includes the direction of wave approach, wave height and the coincidence of storm waves with high tides. Marine processes tend to dominate the majority of morphodynamic conditions until there is a large enough flood event in the associated river to breach the barrier. The level and frequency of base or flood flows are attributed to fluvial processes. Antecedent barrier conditions are the permeability, volume and height of the barrier as well as the width and presence of previous outlet channels. During low to medium river flows, the outlet from the lagoon to the sea becomes offset in the direction of longshore drift. Outlet efficiency tends to decrease the further away from the main river-mouth the outlet is. A decrease in efficiency can cause the outlet to become choked with sediment and the hapua to close temporarily. The potential for closure varies between different hapua depending on whether marine or fluvial processes are the bigger driver in the event. A high flow event; such as a fresh or flood can breach the barrier directly opposite the main river channel. This causes an immediate decrease in the water level of the hapua; as well as transporting previously deposited sediments into the ocean. Flood events are important for eroding lagoon back shores; this is a behaviour which allows hapua to retreat landward and thus remain coastal landforms even with coastal transgression and sea level rise. During high flow events there is also the possibility for secondary breaches of the barrier or lagoon truncation to occur.

Storm events also have the ability to close hapua outlets as waves overtop the barrier depositing sediment and choking the scoured channel. The resultant swift increase in lagoon water level causes a new outlet to be breached rapidly due to the large hydraulic head that forms between the lagoon and sea water levels. Storm breaching is believed to be an important but unpredictable control on the duration of closures at low to moderate river flow levels in smaller hapua.

Hapua are extremely important for a number of reasons. They provide a link between the river and sea for migrating fish as well as a corridor for migratory birds. To lose this link via closure of the hapua outlet could result in losing entire generations of specific species as they may need to migrate to the ocean or the river as a vital part of their lifecycle. River-mouth lagoons such as hapua were also used a source for *mahinga kai* (food gathering) by the Māori people. However, this is no longer the case due to catchment degradation which has resulted in lagoon deterioration. River-mouth lagoons on MSG beaches are not well explained in international literature.

Hapua Case Study

Aerial photograph of the Rakaia river-mouth and associated hapua

The hapua located at the mouth of the Rakaia River stretches approximately three kilometres north from where the river-mouth reaches the coast. The average width of the hapua between 1952 and 2004 was approximately 50 metres; whilst the surface area has stabilised at approximately 600,000 square metres since 1966. The coastal hinterland is composed of erodible cliffs and a low-lying area commonly known as the Rakaia Huts. This area has changed notably since European Settlement; with the drainage of ecologically significant wetlands and development of the small bach community.

The Rakaia River begins in the Southern Alps, providing approximately 4.2 Mt per year of sediment to the east coast. It is a braided river with a catchment area of 3105 kilometres squared and a mean flow of 221 cubic metres per second. The mouth of the Rakaia River reaches the coast south of Banks Peninsula. As the river reaches the coast it diverges into two channels; with the main channel flowing to the south of the island. As the hapua is located in the Canterbury Bight it is in a state of constant morphological change due to the prevailing southerly sea swells and resultant northwards longshore drift.

Mangrove

Mangroves are shrubs or small trees that grow in coastal saline or brackish water. The term is also used for tropical coastal vegetation consisting of such species. Mangroves occur worldwide in the tropics and subtropics, mainly between latitudes 25° N and 25° S. In the year 2000, the area of mangroves was 53,190 square miles (137,760 km²), spanning 118 countries and territories.

Mangroves are salt tolerant trees, also called halophytes, and are adapted to life in harsh coastal conditions. They contain a complex salt filtration system and complex root system to cope with salt water immersion and wave action. They are adapted to the low oxygen (anoxic) conditions of waterlogged mud.

The word is used in at least three senses: (1) most broadly to refer to the habitat and entire plant assemblage or *mangal*, for which the terms *mangrove forest biome*, *mangrove swamp* and *mangrove forest* are also used, (2) to refer to all trees and large shrubs in the mangrove swamp, and (3) narrowly to refer to the mangrove family of plants, the Rhizophoraceae, or even more specifically just to mangrove trees of the genus *Rhizophora*.

The mangrove biome, or mangal, is a distinct saline woodland or shrubland habitat characterized by depositional coastal environments, where fine sediments (often with high organic content) collect in areas protected from high-energy wave action. The saline conditions tolerated by various mangrove species range from brackish water, through pure seawater (3 to 4 %), to water concentrated by evaporation to over twice the salinity of ocean seawater (up to 9 %).

Etymology

The term "mangrove" comes to English from Spanish (perhaps by way of Portuguese), and is likely to originate from Guarani. It was earlier "mangrow" (from Portuguese *mangue* or Spanish *mangle*), but this word was corrupted via folk etymology influence of the word "grove".

Ecology

The world's mangrove forests in 2000.

Mangrove swamps are found in tropical and subtropical tidal areas. Areas where mangal occurs include estuaries and marine shorelines.

The intertidal existence to which these trees are adapted represents the major limitation to the number of species able to thrive in their habitat. High tide brings in salt water, and when the tide recedes, solar evaporation of the seawater in the soil leads to further increases in salinity. The return of tide can flush out these soils, bringing them back to salinity levels comparable to that of seawater.

At low tide, organisms are also exposed to increases in temperature and desiccation, and are then cooled and flooded by the tide. Thus, for a plant to survive in this environment, it must tolerate broad ranges of salinity, temperature, and moisture, as well as a number of other key environmental factors — thus only a select few species make up the mangrove tree community.

About 110 species are considered "mangroves", in the sense of being a tree that grows in such a saline swamp, though only a few are from the mangrove plant genus, *Rhizophora*. However, a given mangrove swamp typically features only a small number of tree species. It is not uncommon for a mangrove forest in the Caribbean to feature only three or four tree species. For comparison, the tropical rainforest biome contains thousands of tree species, but this is not to say mangrove forests lack diversity. Though the trees themselves are few in species, the ecosystem that these trees create provides a home (habitat) for a great variety of other organisms.

Mangrove plants require a number of physiological adaptations to overcome the problems of anoxia, high salinity and frequent tidal inundation. Each species has its own solutions to these problems; this may be the primary reason why, on some shorelines, mangrove tree species show distinct zonation. Small environmental variations within a mangal may lead to greatly differing methods for coping with the environment. Therefore, the mix of species is partly determined by the tolerances of individual species to physical conditions, such as tidal inundation and salinity, but may also be influenced by other factors, such as predation of plant seedlings by crabs.

Once established, mangrove roots provide an oyster habitat and slow water flow, thereby enhancing sediment deposition in areas where it is already occurring. The fine, anoxic sediments under mangroves act as sinks for a variety of heavy (trace) metals which colloidal particles in the sediments have scavenged from the water. Mangrove removal disturbs these underlying sediments, often creating problems of trace metal contamination of seawater and biota.

Mangrove swamps protect coastal areas from erosion, storm surge (especially during hurricanes), and tsunamis. The mangroves' massive root systems are efficient at dissipating wave energy. Likewise, they slow down tidal water enough so its sediment is deposited as the tide comes in, leaving all except fine particles when the tide ebbs. In this way, mangroves build their own environments. Because of the uniqueness of mangrove ecosystems and the protection against erosion they provide, they are often the object of conservation programs, including national biodiversity action plans.

However, mangrove swamps' protective value is sometimes overstated. Wave energy is typically low in areas where mangroves grow, so their effect on erosion can only be measured over long periods. Their capacity to limit high-energy wave erosion is limited to events such as storm surges and tsunamis. Erosion often occurs on the outer sides of bends in river channels that wind through mangroves, while new stands of mangroves are appearing on the inner sides where sediment is accruing.

The unique ecosystem found in the intricate mesh of mangrove roots offers a quiet marine region for young organisms. In areas where roots are permanently submerged, the organisms they host include algae, barnacles, oysters, sponges, and bryozoans, which all require a hard surface for anchoring while they filter feed. Shrimps and mud lobsters use the muddy bottoms as their home. Mangrove crabs munch on the mangrove leaves, adding nutrients to the mangal muds for other bottom feeders. In at least some cases, export of carbon fixed in mangroves is important in coastal food webs.

Mangrove plantations in Vietnam, Thailand, Philippines and India host several commercially important species of fishes and crustaceans. Despite restoration efforts, developers and others have removed over half of the world's mangroves in recent times.

Mangrove forests can decay into peat deposits because of fungal and bacterial processes as well as by the action of termites. It becomes peat in good geochemical, sedimentary and tectonic conditions. The nature of these deposits depends on the environment and the types of mangrove involved. In Puerto Rico the red (Rhizophora mangle), white (Laguncularia racemosa) and black (Avicennia germinans) mangroves occupy different ecological niches and have slightly different chemical compositions so the carbon content varies between the species as well between the different tissues of the plant e.g. leaf matter vs roots.

In Puerto Rico there is a clear succession of these three trees from the lower elevations which are dominated by red mangroves to farther inland with a higher concentration of white mangroves. Mangrove forests are an important part of the cycling and storage of carbon in tropical coastal ecosystems. Using this it is possible to attempt to reconstruct the environment and investigate changes to the coastal ecosystem for thousands of years by using sediment cores. However, an additional complication is the imported marine organic matter that also gets deposited in the sediment due to tidal flushing of mangrove forests.

In order to understand peat formation by mangroves, it is important to understand the conditions they grew in, and how they decayed. Termites are an important part of this decay, and so an understanding of their action on the organic matter is crucial to the chemical stabilization of mangrove peats.

Biology

Of the recognized 110 mangrove species, only about 54 species in 20 genera from 16 families constitute the "true mangroves", species that occur almost exclusively in mangrove habitats. Demonstrating convergent evolution, many of these species found similar solutions to the tropical conditions of variable salinity, tidal range (inundation), anaerobic soils and intense sunlight. Plant biodiversity is generally low in a given mangal. The greatest biodiversity occurs in the mangal of New Guinea, Indonesia and Malaysia.

Adaptations to Low Oxygen

A red mangrove, *Rhizophora mangle*.

Above and below water view at the edge of the mangal.

Red mangroves, which can survive in the most inundated areas, prop themselves above the water level with stilt roots and can then absorb air through pores in their bark (lenticels). Black mangroves live on higher ground and make many pneumatophores (specialised root-like structures which stick up out of the soil like straws for breathing) which are also covered in lenticels.

These "breathing tubes" typically reach heights of up to 30 cm, and in some species, over 3 m. The four types of pneumatophores are stilt or prop type, snorkel or peg type, knee type, and ribbon or plank type. Knee and ribbon types may be combined with buttress roots at the base of the tree. The roots also contain wide aerenchyma to facilitate transport within the plants.

Limiting Salt Intake

Red mangroves exclude salt by having significantly impermeable roots which are highly suberised (impregnated with suberin), acting as an ultra-filtration mechanism to exclude sodium salts from the rest of the plant. Analysis of water inside mangroves has shown 90% to 97% of salt has been excluded at the roots. In a frequently cited concept that has become known as the "sacrificial leaf", salt which does accumulate in the shoot (sprout) then concentrates in old leaves, which the plant then sheds. However, recent research suggests the older, yellowing leaves have no more measurable salt content than the other, greener leaves. Red mangroves can also store salt in cell vacuoles. As seen in the photograph on the right, white or grey mangroves can secrete salts directly; they have two salt glands at each leaf base (correlating with their name—they are covered in white salt crystals).

Salt crystals formed on grey mangrove leaf.

Limiting Water Loss

Because of the limited fresh water available in salty intertidal soils, mangroves limit the amount of water they lose through their leaves. They can restrict the opening of their stomata (pores on the leaf surfaces, which exchange carbon dioxide gas and water vapour during photosynthesis). They also vary the orientation of their leaves to avoid the harsh midday sun and so reduce evaporation from the leaves. Anthony Calfo, a noted aquarium author, observed anecdotally a red mangrove in captivity only grows if its leaves are misted with fresh water several times a week, simulating frequent tropical rainstorms.

Nutrient Uptake

Because the soil is perpetually waterlogged, little free oxygen is available. Anaerobic bacteria liberate nitrogen gas, soluble ferrum (iron), inorganic phosphates, sulfides and methane, which make the soil much less nutritious. Pneumatophores (aerial roots) allow mangroves to absorb gases directly from the atmosphere, and other nutrients such as iron, from the inhospitable soil. Mangroves store gases directly inside the roots, processing them even when the roots are submerged during high tide.

Increasing Survival of Offspring

In this harsh environment, mangroves have evolved a special mechanism to help their offspring survive. Mangrove seeds are buoyant and are therefore suited to water dispersal. Unlike most plants, whose seeds germinate in soil, many mangroves (e.g. red mangrove) are viviparous, whose seeds germinate while still attached to the parent tree. Once germinated, the seedling grows either within the fruit (e.g. *Aegialitis, Avicennia* and *Aegiceras*), or out through the fruit (e.g. *Rhizophora, Ceriops, Bruguiera* and *Nypa*) to form a propagule (a ready-to-go seedling) which can produce its own food via photosynthesis.

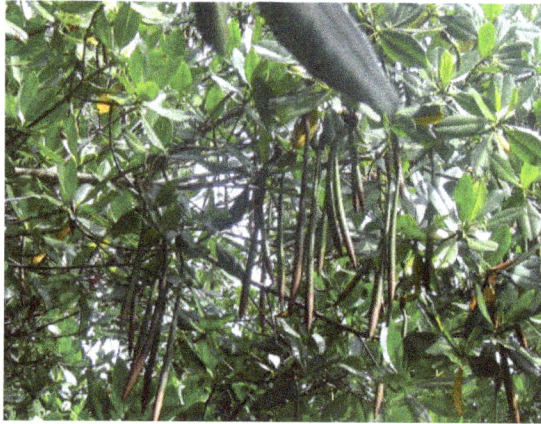

Red mangrove seeds germinate while still on the parent tree.

The mature propagule then drops into the water, which can transport it great distances. Propagules can survive desiccation and remain dormant for over a year before arriving in a suitable environment. Once a propagule is ready to root, its density changes so the elongated shape now floats vertically rather than horizontally. In this position, it is more likely to lodge in the mud and root. If it does not root, it can alter its density and drift again in search of more favorable conditions.

Taxonomy and Evolution

The following listing (modified from Tomlinson, 1986) gives the number of species of mangroves in each listed plant genus and family. Mangrove environments in the Eastern Hemisphere harbor six times as many species of trees and shrubs as do mangroves in the New World. Genetic divergence of mangrove lineages from terrestrial relatives, in combination with fossil evidence, suggests mangrove diversity is limited by evolutionary transition into the stressful marine environment, and the number of mangrove lineages has increased steadily over the Tertiary with little global extinction.

Major Components

Family	Genus, number of species	Common name
Acanthaceae, Avicenniaceae or Verbenaceae (family allocation disputed)	*Avicennia*, 9	Black mangrove
Combretaceae	*Conocarpus*, 1; *Laguncularia*, 11; *Lumnitzera*, 2	Buttonwood, white mangrove
Arecaceae	*Nypa*, 1	Mangrove palm
Rhizophoraceae	*Bruguiera*, 6; *Ceriops*, 2; *Kandelia*, 1; *Rhizophora*, 8	Red mangrove
Lythraceae	*Sonneratia*, 5	Mangrove apple

Minor Components

Family	Genus, number of species
Acanthaceae	*Acanthus*, 1; *Bravaisia*, 2
Bombacaceae	*Camptostemon*, 2
Cyperaceae	*Fimbristylis*, 1
Euphorbiaceae	*Excoecaria*, 2
Lecythidaceae	*Barringtonia*, 6
Lythraceae	*Pemphis*, 2
Meliaceae	*Xylocarpus*, 2
Myrtaceae	*Osbornia*, 1
Pellicieraceae	*Pelliciera*, 1
Plumbaginaceae	*Aegialitis*, 2
Primulaceae	*Aegiceras*, 2
Pteridaceae	*Acrostichum*, 3
Rubiaceae	*Scyphiphora*, 1
Sterculiaceae	*Heritiera*, 3

Geographical Regions

Mangroves can be found in over 118 countries and territories in the tropical and subtropical regions of the world. The largest percentage of mangroves is found between the 5° N and 5° S latitudes. Approximately 75% of world's mangroves are found in just 15 countries. Asia has the largest amount (42%) of the world's mangroves, followed by Africa (21%), North/Central America (15%), Oceania (12%) and South America (11%).

Africa

There are important mangrove swamps in Kenya, Tanzania, République Démocratique du Congo (RDC) and Madagascar, with the latter even admixing at the coastal verge with dry deciduous forests.

Nigeria has Africa's largest mangrove concentration, spanning 36,000 km². Oil spills and leaks have destroyed many in the last 50 years, damaging the local fishing economy and water quality.

Along the coast of the Red Sea, both on the Egyptian side and in the Gulf of Aqaba, mangroves composed primarily of *Avicennia marina* and *Rhizophora mucronata* grow in about 28 stands that cover about 525 hectares. Almost all Egyptian mangrove stands are now protected.

There are mangroves off the coast of Durban, South Africa, on the Umgeni River estuary.

Americas

Mangroves live in many parts of the tropical and subtropical coastal zones of North and South America.

Continental United States

Because of their sensitivity to subfreezing temperatures, mangroves in the continental United States are very limited to the Florida peninsula and some isolated growths of black mangrove (*Avicennia germinans*) at the southmost coast of Louisiana and South Texas.

Mexico

In Mexico four species of mangrove predominate: *Rhizophora mangle, Laguncularia racemosa, Avicennia germinans* and *Conocarpus erectus*. During an inventory conducted by CONABIO between 2006 and 2008, 770,057 hectares of mangrove were counted. Of this total, 55% are located in the Yucatán Peninsula.

Significant mangals include the Marismas Nacionales-San Blas mangroves found in Sinaloa and Nayarit.

Central America and Caribbean

Mangroves occur on the Pacific and Caribbean coasts of Belize, Costa Rica, El Salvador, Guatemala, Honduras, Nicaragua, and Panama. Mangroves can also be found in many of the Antilles including Puerto Rico, Cuba, and Hispaniola, as well as other islands in the West Indies such as the Bahamas.

The same area in Honduras shown from 1987 (bottom) to 1999 and the corresponding removal of mangrove swamps for shrimp farming.

Belize

The nation of Belize has the highest overall percentage of forest cover of any of the Central American countries. In terms of Belize's mangrove cover—which assumes the form not only of mangrove 'forest', but also of scrubs and savannas, among others—a 2010 satellite-based study of Belize's mangroves by the World Wildlife Fund (WWF) and the Water Center for the Humid Tropics of Latin America and the Caribbean found, in 2010, mangroves covered some 184,548 acres (74,684 hectares) or 3.4% of Belize's territory.

In 1980, by contrast, mangrove cover stood at 188,417 acres (76,250 hectares)—also 3.4% of Belize's territory, although based on the work of mangrove researcher Simon Zisman, Belize's mangrove cover in 1980 was estimated to represent 98.7% of the precolonial extent of those ecosystems. Belize's mangrove cover in 2010 was thus estimated to represent 96.7% of the precolonial cover. Assessing changes in Belize's mangrove cover over a 30-year period was possible because of Belize's participation in the Regional Visualization and Monitoring System, a regional observatory jointly implemented by CATHALAC, RCMRD, ICIMOD, NASA, USAID, and other partners.

South America

Brazil contains approximately 26,000 km² of mangals, 15% of the world's total.

Ecuador has substantial remaining mangrove forests in the provinces of El Oro, Guayas, Manabi and Esmeraldas with limited forest remaining in Santa Elena. The northern portion of Esmeraldas province has a large pristine mangrove forest that is preserved as the Reserva Ecológica Cayapas-Mataje (REMACAN) and is an original Ramsar site. This forest is the most preserved within Ecuador and likely the most pristine forest along the Pacific Coast of the Americas.

Recife in Brazil, nicknamed the "Manguetown", has the largest urban mangrove forest in the world.

The only other major mangrove holding in Esmeraldas is in-and-around the community of Muisne and the Rio Muisne Estuary Swampland Wildlife Refuges. The mangroves in-and-around the estuaries of Muisne have decreased in area from 3222 ha in 1971 to 1065 ha as of 2005, during this time commercial shrimp aquaculture has become the dominant land-cover within this estuary environment.

On the border of Esmeraldas province and Manabí province is a formerly large area of mangrove within Cojimies Estuary. The mangroves in this estuary are some of the most degraded in Ecuador with only 19% of 1971 mangrove area remaining as of 1998, although mangrove has recovered since this date. Within Manabí the major mangrove holding estuary is the Chone estuary situated near the city of Bahía de Caráquez. Again, Chone has undergone substantial mangrove deforestation since the advent of commercial aquaculture in Ecuador. Although mangrove loss appears to have halted in this estuary and mangrove regrowth driven by local fisherman is now occurring.

Peru has a very small region of mangrove located in the north-west of the country on the Ecuadorian Border.

Venezuela's northern Caribbean island, Margarita, possesses mangrove forests in the Parque nacional Laguna de La Restinga. Venezuela has 4% of the world's mangroves, with an extension of 6735 km².

Colombia possesses large mangrove forests on both its Caribbean and Pacific coasts.

The Mangrove forests of Suriname have a height of 20–25 m and are found mainly in the coastal area. There are six types of mangroves, namely two types of parwa or black mangroves, three types of red mangroves (mangro) and a small mangrove species (white mangrove, akira or tjila).

Asia

Bangladesh

The Sundarbans, a UNESCO World Heritage Site, is the largest single block of tidal halophytic mangrove forest in the world, covering parts of Bangladesh's Khulna Division and the Indian state of West Bengal.

The Sundarbans National Park is a National Park, Tiger Reserve, and a Biosphere Reserve. It is one of the largest reserves for the Bengal tiger.

A third of this area is covered by water and marshes. Since 1966 it has been considered a sanctuary for wildlife with an estimated 400 Royal Bengal tigers and more than 30,000 deer.

The forest lies at the foot of the Ganges The seasonally-flooded Sundarbans freshwater swamp forests lie inland from the mangrove forests. The forest covers 10,000 square kilometres (3,900 sq mi) of which about 6,000 square kilometres (2,300 sq mi) are in Bangladesh. It was inscribed as a UNESCO world heritage site in 1997, but while the Bangladeshi and Indian portions constitute the same continuous ecotope, they are separately listed in the UNESCO world heritage list as the Sundarbans and Sundarbans National Park, respectively. The Sundarbans is intersected by a complex network of tidal waterways, mudflats and small islands of salt-tolerant mangrove forests. The area is known for the eponymous Royal Bengal Tiger (Panthera tigris tigris), as well as numerous fauna including species of birds, spotted deer, crocodiles and snakes. Sundarbans was designated a Ramsar site on May 21, 1992. The fertile soils of the delta have been subject to intensive human use for centuries, and the ecoregion has been mostly converted to intensive agriculture, with few enclaves of forest remaining. The remaining forests, together with the Sundarbans mangroves, are important habitat for the endangered tiger. Additionally, the Sundarbans serves a crucial function as a protective flood barrier for the millions of inhabitants in and around Kolkata (Calcutta) against the result of cyclone activity. Sundarbans is home to many different species of birds, mammals, insects, reptiles and fish. It is estimated that there may be found more than 120 species of fish and over 260 species of birds and more than fifty species of reptiles and eight amphibians. Many tourists go there to see the Bengal tigers, saltwater crocodiles, leopards and snakes cobra. Geographical data of Sundarban Latitude: 25.7667 Longitude: 88.7167 Average Height: 34 Time Zone: Asia / Dhaka Title: Lugar Poblado.

India

The deltas of the Ganges, Mahanadi, Krishna, Godavari, and Kaveri rivers are known to contain mangrove forests.

The following table shows the presence of mangroves in the different states of India and the total area covered by them in square kilometres.

Rank	States/UTs with Highest Mangrove Cover 2013	Total Mangrove Cover in km²
1	West Bengal	2,097
2	Gujarat	1,103
3	Andaman And Nicobar Islands	604
4	Andhra Pradesh	352
5	Odisha	213
6	Maharashtra	186
7	Tamil Nadu	39
8	Goa	22
9	Kerala	6
10	Karnataka	3

Sundarban Mangroves (India & Bangladesh)

In the Ganga-Brahmaputra delta, Sundari trees are found, which provide durable hard timber. Palm, Coconut, keora, agar, also grow in some parts of the delta. India's mangrove forests are known to serve as a habitat for turtles, crocodiles, gharials, and snakes. This region is part of the Great Sundarbans, the largest mangroves region in the world and a UNESCO World Heritage Site. This area covers a National Park, Tiger reserve and a Biosphere Reserve.

Bhitarkanika Mangroves

Bhitarkanika Mangroves is India's second largest forest, located in the state of Odisha. Bhitarkanika is created by the two river deltas of Brahmani and Baitarani river and one of the important Ramsar Wetland in India. It is also home of salt-water crocodiles and nesting beach olive ridley sea turtle.

Godavari-Krishna Mangroves

The Godavari-Krishna mangroves lies in the delta of the Godavari and Krishna rivers in the state of Andhra Pradesh. Mangroves ecoregion is under protection for Calimere Wildlife and Pulicat Lake Bird Sanctuary.

Pichavaram Mangroves

Pichavaram mangrove is one of the largest mangrove in India, situated at Pichavaram near Chidambaram in the state of Tamil Nadu. Pichavaram ranks amongs one of the most exquisite scenic spots in Tamil Nadu and home of many species of aquatic birds.

Baratang Island Mangroves

Baratang Island Mangroves is a swamp, located at Great Andaman and Nicobar Islands. Mangrove Swamps of Baratang Island are situated between Middle and South Andamans, capital city Port Blair.

Sunderbans

Indo-malaya Ecozone

Mangroves occur on Asia's south coast, throughout the Indian subcontinent, in all Southeast Asian countries, and on islands in the Indian Ocean, Persian gulf, Arabian Sea, Bay of Bengal, South China Sea, East China Sea and the Pacific.

The mangal is particularly prevalent in the deltas of large Asian rivers. The Sundarbans is the largest mangrove forest in the world, located in the Ganges River delta in Bangladesh and West Bengal, India.

The Pichavaram mangroves in Tamil Nadu is India's one of the largest mangrove forests. The Bhitarkanika Mangroves Forest of Odisha, by the Bay of Bengal, is India's second largest mangrove forest. Other major mangals occur on the Andaman and Nicobar Islands and the Gulf of Kutch in Gujarat.

Mangroves occur in certain muddy swampy islands of the Maldives.

On the Malayan Peninsula mangroves cover an estimated 1,089.7 square kilometres (420.7 sq mi), while most of the remaining 5,320 square kilometres (2,054 sq mi) mangroves in Malaysia are on the island of Borneo.

In Vietnam, mangrove forests grow along the southern coast, including two forests: the Can Gio Mangrove Forest biosphere reserve and the U Minh mangrove forest in the sea and coastal region of Kiên Giang, Cà Mau and Bạc Liêu provinces.

The mangrove forests of Kompong Sammaki in Cambodia are of major ecological and cultural importance, as the human population relies heavily on the crabs and fish that live in the roots.

The three most important mangrove forests of Taiwan are: Tamsui River in Taipei, Jhonggang River in Miaoli and the Sihcao Wetlands in Tainan. According to research, four types of mangrove exist in Taiwan. Some places have been developed as scenic areas, such as the log raft routes in Sihcao.

The most extensive mangrove forests of the Ryukyu Islands in East China Sea occur on Iriomote Island of the Yaeyama Islands, Okinawa, Japan. Seven types of mangroves are recognised on Iriomote Island.

The northern limit of mangrove forests in the Indo-Malaya ecozone is considered to be Tanegashima Island, Kyushu, Japan.

Indonesia

In the Indonesian Archipelago, mangroves occur around much of Sumatra, Borneo, Sulawesi, and the surrounding islands, while further north, they are found along the coast of the Malay Peninsula. Indonesia has around 9.36 million hectares of mangrove forests, but 48% is categorized as 'moderately damaged' and 23% as 'badly damaged'.

GUINNESS WORLD RECORDS

CERTIFICATE

The most trees planted in 24 hours by a team of 300 is 847,275 and was achieved by the Sindh Forest Department at Kharochan, Distrct Thatta, Pakistan, on June 22 2013

OFFICIALLY AMAZING

The Guinness World Record certificate of achievement to Sindh Forest Departmet, Govt of Sindh, Pakistan

Mangroves Around the World

Sihcao, Tainan, Taiwan

Glowing mangrove plantation at Keti Bundar, Thatta, Pakistan

Flourishing mangroves along Karachi coast, Pakistan

Pakistan

Pakistani mangroves are located mainly along the delta of the Indus River (the Indus River Delta-Arabian Sea mangroves ecoregion). Major mangrove forests are found on the coastline of the provinces of Sindh and Balochistan. In Karachi, land reclamation projects have led to the cutting down of mangrove forests for commercial and urban development.. On 22 June 2013, Sindh Forest Department, Govt. of Sindh, Pakistan, with the help of 300 local coastal volunteer planters set the Guinness World through 847,250 mangrove saplings planted at Kharo Chan, Thatta, Sindh, Pakistan in a little over 12 hours. This is the highest number of saplings planted within a day under the Guinness World Record category of *"Maximum Number of Trees Planted in a Day"*..

Shah Bundar, Sujawal, Pakistan, new mangrove plantation

Sindh Forest Department, Government of Sindh Mangrove has played pioneer role for the conservation and protection of the mangroves in the Indus Delta since late 1950s when it was handed over the areas. A breakthrough success is the Re-introduction of Rhizophora mucronata into the Indus Delta, which had become extinct there. More recently, a threatened mangrove shrub Ceriops tagal has also been successfully Re-introduces. A third species, Aegiceras corniculatum, is under trails at the nursery stage.

A gigantic initiative is under in the Sindh, Pakistan, to rehabilitate the degraded and blank mangrove mudflats. Since 2010 alone, around 55,000 Hectares of such area has been planted and rehabilitated. During this period, through concerted efforts and a rigorous conservation policy adopted by the Sindh Forest Department, Govt. of Sindh and the federal govt. a mangrove resource base of 150,000 plus Hectares has been created, with the support of local coastal communities. International organizations like IUCN and WWF are also playing critical role to support this initiative of the government. Other achievements include: (1) Declaring all the mangrove forests in the Indus Delta as Protected Forests in December 2010; Constitution of a Mangrove Conservation Committee at the provincial level which includes all stakeholders as members and overall awareness of the importance of mangroves and its ecosystem.

Middle East

Oman, near Muscat, supports large areas of mangroves, in particular at Shinas, Qurm Park and Mahout Island. In Arabic, mangrove trees are known as *qurm*, thus the mangrove area in Oman is known as Qurm Park. A small mangrove area is present in the Kingdom of Bahrain. Mangroves are also present extensively in neighbouring Yemen.

Iranian mangrove forests occur between 25°11′N to 27°52′N. These forests exist in the north part of the Persian Gulf and Sea of Oman, along three maritime provinces in the south of Iran. These provinces, respectively, from southwest to southeast of Iran, include Bushehr, Hormozgan, and Sistan and Balouchestan.

Mangrove is also widely seen in Tarut Island, east of Qatif in Saudi Arabia. In addition, large forest of mangrove surround the coast to the south of Qatif (Siahat Beach). Nonetheless, because of sea land re-claiming the mangrove is being cut down which makes lots of sea fish losses their natural habitats.

The mangrove forests that cover thousands of hectares of land along the UAE shoreline form an integral part of its coastal ecosystem. The Environment Agency – Abu Dhabi (EAD) is currently working on rehabilitation, conservation and protection of mangrove forests in seven key sites in Abu Dhabi including: Saadiyat Island, Jubail Island, Marawah Marine Biosphere Reserve (which also comprises famous Bu Tinah Island), Bu Syayeef Protected Area, Ras Gharab, the Eastern Corniche and Ras Ghanada.

Oceania

Australia and New Guinea

More than 5 species of Rhizophoraceae grow in Australasia with particularly high biodiversity on the island of New Guinea and northern Australia.

Australia has about 11,500 km² of mangroves, primarily on the northern and eastern coasts of the continent, with occurrences as far south as Millers Landing in Wilsons Promontory, Victoria (38°54′S) and Barker Inlet in Adelaide, South Australia.

New Zealand

New Zealand also has mangrove forests extending to around 38°S (similar to Australia's southernmost mangrove incidence): the furthest geographical extent on the west coast is Raglan Harbour (37°48′S); on the east coast, Ohiwa Harbour (near Opotiki) is the furthest south mangroves are found (38°00′S).

Pacific Islands

Twenty-five species of mangrove are found on various Pacific islands, with extensive mangals on some islands. Mangals on Guam, Palau, Kosrae and Yap have been badly affected by development.

Mangroves are not native to Hawaii, but the red mangrove, *Rhizophora mangle*, and Oriental mangrove, *Bruguiera sexangula*, have been introduced and are now naturalized. Both species are considered invasive species and classified as pests by the University of Hawaii Botany Department.

Exploitation and Conservation

Adequate data is only available for about half of the global area of Mangrove swamp. However, of those areas for which data has been collected, it appears that 35% of the Mangroves have been destroyed. The United Nations Environment Program & Hamilton (2013), estimate that shrimp farming causes approximately a quarter of the destruction of mangrove forests. Likewise, the 2010 update of the World Mangrove Atlas indicated a fifth of the world's mangrove ecosystems have been lost since 1980.

Mangroves in West Bali National Park, Indonesia.

Grassroots efforts to save mangroves from development are becoming more popular as their benefits become more widely known. In the Bahamas, for example, active efforts to save mangroves are occurring on the islands of Bimini and Great Guana Cay. In Trinidad and Tobago as well, efforts are underway to protect a mangrove threatened by the construction of a steelmill and a port. In Thailand, community management has been effective in restoring damaged mangroves. Within northern Ecuador mangrove regrowth is reported in almost all estuaries and stems primarily from local actors responding to earlier periods of deforestation in the Esmeraldas region.

Mangroves have been reported to be able to help buffer against tsunami, cyclones, and other storms. One village in Tamil Nadu was protected from tsunami destruction—the villagers in Naluvedapathy planted 80,244 saplings to get into the Guinness Book of World Records. This created a kilometre-wide belt of trees of various varieties. When the tsunami struck, much of the land around the village was flooded, but the village itself suffered minimal damage.

Reforestation

In some areas, mangrove reforestation and mangrove restoration is also underway. Red mangroves are the most common choice for cultivation, used particularly in marine aquariums in a sump to reduce nitrates and other nutrients in the water. Mangroves also appear in home aquariums, and as ornamental plants, such as in Japan.

In Senegal, Haïdar El Ali has started the fr project, which (amongst others) focuses on reforesting several areas with mangroves.

Mangroves in Bohol, Philippines.

The Manzanar Mangrove Initiative is an ongoing experiment in Arkiko, Eritrea, part of the Manzanar Project founded by Gordon H. Sato, establishing new mangrove plantations on the coastal mudflats. Initial plantings failed, but observation of the areas where mangroves did survive by themselves led to the conclusion that nutrients in water flow from inland were important to the health of the mangroves. Trials with the Eritrean Ministry of Fisheries followed, and a planting system was designed to provide the nitrogen, phosphorus, and iron missing from seawater.

The propagules are planted inside a reused galvanized steel can with the bottom knocked out; a small piece of iron and a pierced plastic bag with fertilizer containing nitrogen and phosphorus are buried with the propagule. As of 2007[update], after six years of planting, 700,000 mangroves are growing; providing stock feed for sheep and habitat for oysters, crabs, other bivalves, and fish.

National Studies

In terms of local and national studies of mangrove loss, the case of Belize's mangroves is illustrative in its contrast to the global picture. A recent, satellite-based study—funded by the World Wildlife Fund and conducted by the Water Center for the Humid Tropics of Latin America and the Caribbean (CATHALAC)—indicates Belize's mangrove cover declined by a mere 2% over a 30-year period. The study was born out of the need to verify the popular conception that mangrove clearing in Belize was rampant.

Instead, the assessment showed, between 1980 and 2010, under 4,000 acres (16 km^2) of mangroves had been cleared, although clearing of mangroves near Belize's main coastal settlements (e.g. Belize City and San Pedro) was relatively high. The rate of loss of Belize's mangroves—at 0.07% per year between 1980 and 2010—was much lower than Belize's overall rate of forest clearing (0.6% per year in the same period). These findings can also be interpreted to indicate Belize's mangrove regulations (under the nation's) have largely been effective. Nevertheless, the need to protect Belize's mangroves is imperative, as a 2009 study by the World Resources Institute (WRI) indicates the ecosystems contribute US$174–249 million per year to Belize's national economy.

Coral Reef

Coral reefs are diverse underwater ecosystems held together by calcium carbonate structures secreted by corals. Coral reefs are built by colonies of tiny animals found in marine waters that contain few nutrients. Most coral reefs are built from stony corals, which in turn consist of polyps that cluster in groups. The polyps belong to a group of animals known as Cnidaria, which also includes sea anemones and jellyfish. Unlike sea anemones, corals secrete hard carbonate exoskeletons which support and protect the coral polyps. Most reefs grow best in warm, shallow, clear, sunny and agitated waters.

Often called "rainforests of the sea", shallow coral reefs form some of the most diverse ecosystems on Earth. They occupy less than 0.1% of the world's ocean surface, about half the area of France, yet they provide a home for at least 25% of all marine species, including fish, mollusks, worms, crustaceans, echinoderms, sponges, tunicates and other cnidarians. Paradoxically, coral reefs flourish even though they are surrounded by ocean waters that provide few nutrients. They are most commonly found at shallow depths in tropical waters, but deep water and cold water corals also exist on smaller scales in other areas.

Coral reefs deliver ecosystem services to tourism, fisheries and shoreline protection. The annual global economic value of coral reefs is estimated between US$29.8-375 billion. However, coral reefs are fragile ecosystems, partly because they are very sensitive to water temperature. They are under threat from climate change, oceanic acidification, blast fishing, cyanide fishing for aquarium fish, sunscreen use, overuse of reef resources, and harmful land-use practices, including urban and agricultural runoff and water pollution, which can harm reefs by encouraging excess algal growth.

Formation

Most of the coral reefs we can see today were formed after the last glacial period when melting ice caused the sea level to rise and flood the continental shelves. This means that most modern coral reefs are less than 10,000 years old. As communities established themselves on the shelves, the reefs grew upwards, pacing rising sea levels. Reefs that rose too slowly could become drowned reefs. They are covered by so much water that there was insufficient light. Coral reefs are found in the deep sea away from continental shelves, around oceanic islands and as atolls. The vast majority of these islands are volcanic in origin. The few exceptions have tectonic origins where plate movements have lifted the deep ocean floor on the surface.

In 1842 in his first monograph, *The Structure and Distribution of Coral Reefs*, Charles Darwin set out his theory of the formation of atoll reefs, an idea he conceived during the voyage of the *Beagle*. He theorized uplift and subsidence of the Earth's crust under the oceans formed the atolls. Darwin's theory sets out a sequence of three stages in atoll formation. It starts with a fringing reef forming around an extinct volcanic island as the island and ocean floor subsides. As the subsidence continues, the fringing reef becomes a barrier reef, and ultimately an atoll reef.

Darwin's theory starts with a volcanic island which becomes extinct

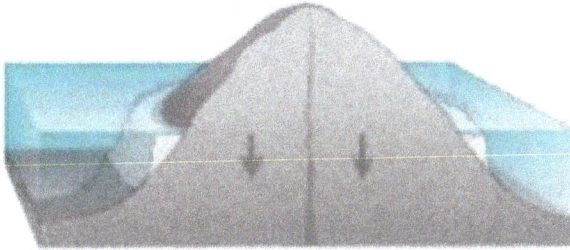

As the island and ocean floor subside, coral growth builds a fringing reef, often including a shallow lagoon between the land and the main reef.

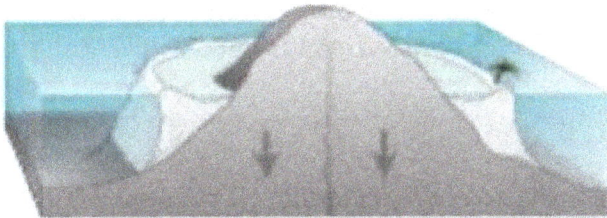

As the subsidence continues, the fringing reef becomes a larger barrier reef further from the shore with a bigger and deeper lagoon inside.

Ultimately, the island sinks below the sea, and the barrier reef becomes an atoll enclosing an open lagoon.

Darwin predicted that underneath each lagoon would be a bed rock base, the remains of the original volcano. Subsequent drilling proved this correct. Darwin's theory followed from his understanding that coral polyps thrive in the clean seas of the tropics where the water is agitated, but can only live within a limited depth range, starting just below low tide. Where the level of the underlying earth allows, the corals grow around the coast to form what he called fringing reefs, and can eventually grow out from the shore to become a barrier reef.

A fringing reef can take ten thousand years to form, and an atoll can take up to 30 million years.

Where the bottom is rising, fringing reefs can grow around the coast, but coral raised above sea level dies and becomes white limestone. If the land subsides slowly, the fringing reefs keep pace by growing upwards on a base of older, dead coral, forming a barrier reef enclosing a lagoon between the reef and the land. A barrier reef can encircle an island, and once the island sinks below sea level a roughly circular atoll of growing coral continues to keep up with the sea level, forming a central lagoon. Barrier reefs and atolls do not usually form complete circles, but are broken in places by storms. Like sea level rise, a rapidly subsiding bottom can overwhelm coral growth, killing the coral polyps and the reef, due to what is called *coral drowning*. Corals that rely on zooxanthellae can *drown* when the water becomes too deep for their symbionts to adequately photosynthesize, due to decreased light exposure.

The two main variables determining the geomorphology, or shape, of coral reefs are the nature of the underlying substrate on which they rest, and the history of the change in sea level relative to that substrate.

The approximately 20,000-year-old Great Barrier Reef offers an example of how coral reefs formed on continental shelves. Sea level was then 120 m (390 ft) lower than in the 21st century. As sea level rose, the water and the corals encroached on what had been hills of the Australian coastal plain. By 13,000 years ago, sea level had risen to 60 m (200 ft) lower than at present, and many hills of the coastal plains had become continental islands. As the sea level rise continued, water topped most of the continental islands. The corals could then overgrow the hills, forming the present cays and reefs. Sea level on the Great Barrier Reef has not changed significantly in the last 6,000 years, and the age of the modern living reef structure is estimated to be between 6,000 and 8,000 years. Although the Great Barrier Reef formed along a continental shelf, and not around a volcanic island, Darwin's principles apply. Development stopped at the barrier reef stage, since Australia is not about to submerge. It formed the world's largest barrier reef, 300–1,000 m (980–3,280 ft) from shore, stretching for 2,000 km (1,200 mi).

Healthy tropical coral reefs grow horizontally from 1 to 3 cm (0.39 to 1.18 in) per year, and grow vertically anywhere from 1 to 25 cm (0.39 to 9.84 in) per year; however, they grow only at depths shallower than 150 m (490 ft) because of their need for sunlight, and cannot grow above sea level.

Materials

As the name implies, the bulk of coral reefs is made up of coral skeletons from mostly intact coral colonies. As other chemical elements present in corals become incorporated into the calcium carbonate deposits, aragonite is formed. However, shell fragments and the remains of calcareous algae such as the green-segmented genus *Halimeda* can add to the reef's ability to withstand damage from storms and other threats. Such mixtures are visible in structures such as Eniwetok Atoll.

Types

The three principal reef types are:

- Fringing reef – directly attached to a shore, or borders it with an intervening shallow channel or lagoon

- Barrier reef – reef separated from a mainland or island shore by a deep channel or lagoon

- Atoll reef – more or less circular or continuous barrier reef extends all the way around a lagoon without a central island

A small atoll in the Maldives

Inhabited cay in the Maldives

Other reef types or variants are:

- Patch reef – common, isolated, comparatively small reef outcrop, usually within a lagoon or embayment, often circular and surrounded by sand or sea-grass

- Apron reef – short reef resembling a fringing reef, but more sloped; extending out and downward from a point or peninsular shore

- Bank reef – linear or semicircular shaped-outline, larger than a patch reef

- Ribbon reef – long, narrow, possibly winding reef, usually associated with an atoll lagoon

- Table reef – isolated reef, approaching an atoll type, but without a lagoon

- Habili – reef specific to the Red Sea; does not reach the surface near enough to cause visible surf; may be a hazard to ships (from the Arabic for "unborn")

- Microatoll – community of species of corals; vertical growth limited by average tidal height; growth morphologies offer a low-resolution record of patterns of sea level change; fossilized remains can be dated using radioactive carbon dating and have been used to reconstruct Holocene sea levels

- Cays – small, low-elevation, sandy islands formed on the surface of coral reefs from eroded material that piles up, forming an area above sea level; can be stabilized by plants to become habitable; occur in tropical environments throughout the Pacific, Atlantic and Indian Oceans (including the Caribbean and on the Great Barrier Reef and Belize Barrier Reef), where they provide habitable and agricultural land

- Seamount or guyot – formed when a coral reef on a volcanic island subsides; tops of seamounts are rounded and guyots are flat; flat tops of guyots, or *tablemounts*, are due to erosion by waves, winds, and atmospheric processes

Zones

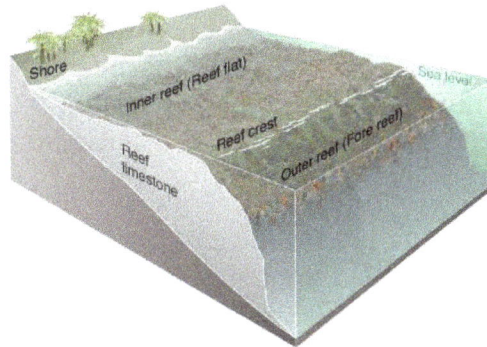

The three major zones of a coral reef: the fore reef, reef crest, and the back reef

Coral reef ecosystems contain distinct zones that represent different kinds of habitats. Usually, three major zones are recognized: the fore reef, reef crest, and the back reef (frequently referred to as the reef lagoon).

All three zones are physically and ecologically interconnected. Reef life and oceanic processes create opportunities for exchange of seawater, sediments, nutrients, and marine life among one another.

Thus, they are integrated components of the coral reef ecosystem, each playing a role in the support of the reefs' abundant and diverse fish assemblages.

Most coral reefs exist in shallow waters less than 50 m deep. Some inhabit tropical continental shelves where cool, nutrient rich upwelling does not occur, such as Great Barrier Reef. Others are found in the deep ocean surrounding islands or as atolls, such as in the Maldives. The reefs surrounding islands form when islands subside into the ocean, and atolls form when an island subsides below the surface of the sea.

Alternatively, Moyle and Cech distinguish six zones, though most reefs possess only some of the zones.

Water in the reef surface zone is often agitated. This diagram represents a reef on a continental shelf. The water waves at the left travel over the *off-reef floor* until they encounter the *reef slope* or *fore reef*. Then the waves pass over the shallow *reef crest*. When a wave enters shallow water it shoals, that is, it slows down and the wave height increases.

The reef surface is the shallowest part of the reef. It is subject to the surge and the rise and fall of tides. When waves pass over shallow areas, they shoal, as shown in the diagram at the right. This means the water is often agitated. These are the precise condition under which corals flourish. Shallowness means there is plenty of light for photosynthesis by the symbiotic zooxanthellae, and agitated water promotes the ability of coral to feed on plankton. However, other organisms must be able to withstand the robust conditions to flourish in this zone.

The off-reef floor is the shallow sea floor surrounding a reef. This zone occurs by reefs on continental shelves. Reefs around tropical islands and atolls drop abruptly to great depths, and do not have a floor. Usually sandy, the floor often supports seagrass meadows which are important foraging areas for reef fish.

The reef drop-off is, for its first 50 m, habitat for many reef fish who find shelter on the cliff face and plankton in the water nearby. The drop-off zone applies mainly to the reefs surrounding oceanic islands and atolls.

The reef face is the zone above the reef floor or the reef drop-off. This zone is often the most diverse area of the reef. Coral and calcareous algae growths provide complex habitats and areas which offer protection, such as cracks and crevices. Invertebrates and epiphytic algae provide much of the food for other organisms.

The reef flat is the sandy-bottomed flat, which can be behind the main reef, containing chunks of coral. This zone may border a lagoon and serve as a protective area, or it may lie between the reef and the shore, and in this case is a flat, rocky area. Fishes tend to prefer living in that flat, rocky area, compared to any other zone, when it is present.

The reef lagoon is an entirely enclosed region, which creates an area less affected by wave action that often contains small reef patches.

However, the "topography of coral reefs is constantly changing. Each reef is made up of irregular patches of algae, sessile invertebrates, and bare rock and sand. The size, shape and relative abundance of these patches changes from year to year in response to the various factors that favor one type of patch over another. Growing coral, for example, produces constant change in the fine structure of reefs. On a larger scale, tropical storms may knock out large sections of reef and cause boulders on sandy areas to move."

Locations

Coral reefs are estimated to cover 284,300 km^2 (109,800 sq mi), just under 0.1% of the oceans' surface area. The Indo-Pacific region (including the Red Sea, Indian Ocean, Southeast Asia and the Pacific) account for 91.9% of this total. Southeast Asia accounts for 32.3% of that figure, while the Pacific including Australia accounts for 40.8%. Atlantic and Caribbean coral reefs account for 7.6%.

Locations of coral reefs

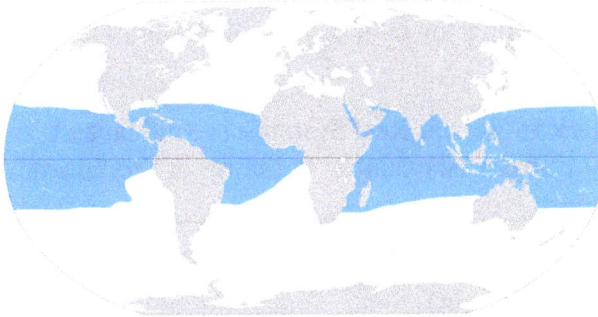

Boundary for 20 °C isotherms. Most corals live within this boundary. Note the cooler waters caused by upwelling on the southwest coast of Africa and off the coast of Peru.

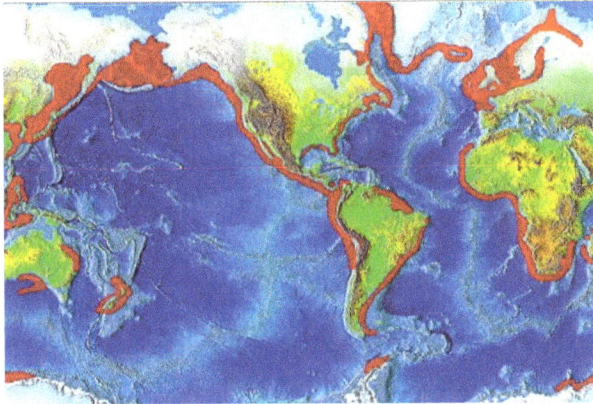

This map shows areas of upwelling in red. Coral reefs are not found in coastal areas where colder and nutrient-rich upwellings occur.

Although corals exist both in temperate and tropical waters, shallow-water reefs form only in a zone extending from approximately 30° N to 30° S of the equator. Tropical corals do not grow at depths of over 50 meters (160 ft). The optimum temperature for most coral reefs is 26–27 °C (79–81 °F), and few reefs exist in waters below 18 °C (64 °F). However, reefs in the Persian Gulf have adapted to temperatures of 13 °C (55 °F) in winter and 38 °C (100 °F) in summer. There are 37 species of scleractinian corals identified in such harsh environment around Larak Island.

Deep-water coral can exist at greater depths and colder temperatures at much higher latitudes, as far north as Norway. Although deep water corals can form reefs, very little is known about them.

Coral reefs are rare along the west coasts of the Americas and Africa, due primarily to upwelling and strong cold coastal currents that reduce water temperatures in these areas (respectively the Peru, Benguela and Canary streams). Corals are seldom found along the coastline of South Asia—from the eastern tip of India (Chennai) to the Bangladesh and Myanmar borders—as well as along the coasts of northeastern South America and Bangladesh, due to the freshwater release from the Amazon and Ganges Rivers respectively.

- The Great Barrier Reef—largest, comprising over 2,900 individual reefs and 900 islands stretching for over 2,600 kilometers (1,600 mi) off Queensland, Australia

- The Mesoamerican Barrier Reef System—second largest, stretching 1,000 kilometers (620 mi) from Isla Contoy at the tip of the Yucatán Peninsula down to the Bay Islands of Honduras

- The New Caledonia Barrier Reef—second longest double barrier reef, covering 1,500 kilometers (930 mi)

- The Andros, Bahamas Barrier Reef—third largest, following the east coast of Andros Island, Bahamas, between Andros and Nassau

- The Red Sea—includes 6000-year-old fringing reefs located around a 2,000 km (1,240 mi) coastline

- The Florida Reef Tract—largest continental US reef, extends from Soldier Key, located in Biscayne Bay, to the Dry Tortugas in the Gulf of Mexico

- Pulley Ridge—deepest photosynthetic coral reef, Florida

- Numerous reefs scattered over the Maldives

- The Philippines coral reef area, the second largest in Southeast Asia, is estimated at 26,000 square kilometers and holds an extraordinary diversity of species. Scientists have identified 915 reef fish species and more than 400 scleractinian coral species, 12 of which are endemic.

- The Raja Ampat Islands in Indonesia's West Papua province offer the highest known marine diversity.

- Bermuda is known for its northernmost coral reef system, located at 32.4° N and 64.8° W. The presence of coral reefs at this high latitude is due to the proximity of the Gulf Stream. Bermuda has a fairly consistent diversity of coral species, representing a subset of those found in the greater Caribbean.

- The world's northernmost individual coral reef so far discovered is located within a bay of Japan's Tsushima Island in the Korea Strait.

- The world's southernmost coral reef is at Lord Howe Island, in the Pacific Ocean off the east coast of Australia.

Biology

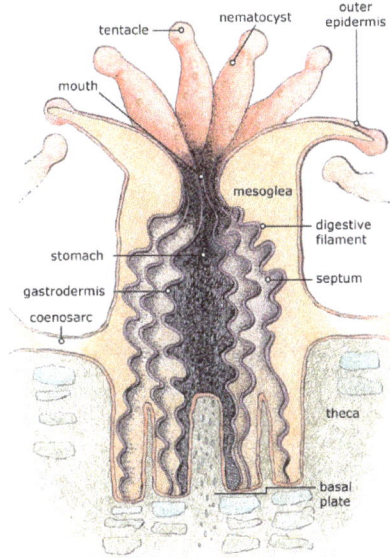

Anatomy of a coral polyp

Alive corals are colonies of small animals embedded in calcium carbonate shells. It is a mistake to think of coral as plants or rocks. Coral heads consist of accumulations of individual animals called polyps, arranged in diverse shapes. Polyps are usually tiny, but they can range in size from a pinhead to 12 inches (30 cm) across.

Reef-building or hermatypic corals live only in the photic zone (above 50 m), the depth to which sufficient sunlight penetrates the water, allowing photosynthesis to occur. Coral polyps do not photosynthesize, but have a symbiotic relationship with micro-scopic algae of the genus *Symbiodinium*, commonly referred to as zooxanthellae. These organisms live within the tissues of polyps and provide organic nutrients that nourish the polyp. Because of this relationship, coral reefs grow much faster in clear water, which admits more sunlight. Without their symbionts, coral growth would be too slow to form significant reef structures. Corals get up to 90% of their nutrients from their symbionts.

Reefs grow as polyps and other organisms deposit calcium carbonate, the basis of coral, as a skeletal structure beneath and around themselves, pushing the coral head's top upwards and outwards. Waves, grazing fish (such as parrotfish), sea urchins, sponges, and other forces and organisms act as bioeroders, breaking down coral skeletons into fragments that settle into spaces in the reef structure or form sandy bottoms in associated reef lagoons. Many other organisms living in the reef community contribute skeletal calcium carbonate in the same manner. Coralline algae are important contributors to reef structure in those parts of the reef subjected to the greatest forces by waves (such as the reef front facing the open ocean). These algae strengthen the reef structure by depositing limestone in sheets over the reef surface.

Typical shapes for coral species are wrinkled brains, cabbages, table tops, antlers, wire strands and pillars. These shapes can depend on the life history of the coral, like light exposure and wave action, and events such as breakages.

Table coral

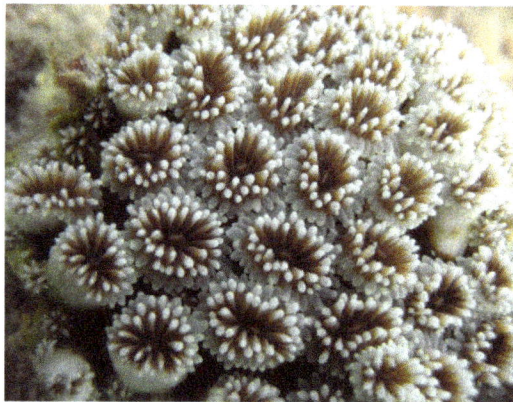

Close up of polyps are arrayed on a coral, waving their tentacles. There can be thousands of polyps on a single coral branch.

Corals reproduce both sexually and asexually. An individual polyp uses both reproductive modes within its lifetime. Corals reproduce sexually by either internal or external fertilization. The reproductive cells are found on the mesenteries, membranes that radiate inward from the layer of tissue that lines the stomach cavity. Some mature adult corals are hermaphroditic; others are exclusively male or female. A few species change sex as they grow.

Internally fertilized eggs develop in the polyp for a period ranging from days to weeks. Subsequent development produces a tiny larva, known as a planula. Externally fertilized eggs develop during synchronized spawning. Polyps release eggs and sperm into the water en masse, simultaneously. Eggs disperse over a large area. The timing of spawning depends on time of year, water temperature, and tidal and lunar cycles. Spawning is most successful when there is little variation between high and low tide. The less water movement, the better the chance for fertilization. Ideal timing occurs in the spring. Release of eggs or planula usually occurs at night, and is sometimes in phase with the lunar cycle (three to six days after a full moon). The period from release to settlement lasts only a few days, but some planulae can survive afloat for several weeks. They are vulnerable to predation and environmental conditions. The lucky few planulae which successfully attach to substrate next confront competition for food and space.

There are eight clades of Symbiodinium phylotypes. Most research has been completed on the Symbiodinium clades A–D. Each one of the eight contributes their own benefits as well as less compatible attributes to the survival of their coral hosts. Each photosynthetic organism has a specific level of sensitivity to photodamage of compounds needed for survival, such as proteins. Rates of regeneration and replication determine the organism's ability to survive. Phylotype A is found more in the shallow regions of marine waters. It is able to produce mycosporine-like amino acids that are UV resistant, using a derivative of glycerin to absorb the UV radiation and allowing them to become more receptive to warmer water temperatures. In the event of UV or thermal damage, if and when repair occurs, it will increase the likelihood of survival of the host and symbiont. This leads to the idea that, evolutionarily, clade A is more UV resistant and thermally resistant than the other clades.

Clades B and C are found more frequently in the deeper water regions, which may explain the higher susceptibility to increased temperatures. Terrestrial plants that receive less sunlight because they are found in the undergrowth can be analogized to clades B, C, and D. Since clades B through D are found at deeper depths, they require an elevated light absorption rate to be able to synthesize as much energy. With elevated absorption rates at UV wavelengths, the deeper occurring phylotypes are more prone to coral bleaching versus the more shallow clades. Clade D has been observed to be high temperature-tolerant, and as a result it has a higher rate of survival than clades B and C.

Brain coral

Staghorn coral

Spiral wire coral

Pillar coral

Darwin's Paradox

"Coral... seems to proliferate when ocean waters are warm, poor, clear and agitated, a fact which Darwin had already noted when he passed through Tahiti in 1842. This constitutes a fundamental paradox, shown quantitatively by the apparent impossibility of balancing input and output of the nutritive elements which control the coral polyp metabolism.

Recent oceanographic research has brought to light the reality of this paradox by confirming that the oligotrophy of the ocean euphotic zone persists right up to the swell-battered reef crest. When you approach the reef edges and atolls from the quasidesert of the open sea, the near absence of living matter suddenly becomes a plethora of life, without transition. So why is there something rather than nothing, and more precisely, where do the necessary nutrients for the functioning of this extraordinary coral reef machine come from?" — Francis Rougerie

In *The Structure and Distribution of Coral Reefs*, published in 1842, Darwin described how coral reefs were found in some areas of the tropical seas but not others, with no obvious cause. The largest and strongest corals grew in parts of the reef exposed to the most violent surf and corals were weakened or absent where loose sediment accumulated.

Tropical waters contain few nutrients yet a coral reef can flourish like an "oasis in the desert". This has given rise to the ecosystem conundrum, sometimes called "Darwin's paradox": "How can such high production flourish in such nutrient poor conditions?"

Coral reefs cover less than 0.1% of the surface of the world's ocean, about half the land area of France, yet they support over one-quarter of all marine species. This diversity results in complex food webs, with large predator fish eating smaller forage fish that eat yet smaller zooplankton and so on. However, all food webs eventually depend on plants, which are the primary producers. Coral reefs' primary productivity is very high, typically producing 5–10 grams of carbon per square meter per day ($gC·m^{-2}·day^{-1}$) biomass.

One reason for the unusual clarity of tropical waters is they are deficient in nutrients and drifting plankton. Further, the sun shines year round in the tropics, warming the surface layer, making it less dense than subsurface layers. The warmer water is separated from deeper, cooler water by a stable thermocline, where the temperature makes a rapid change. This keeps the warm surface waters floating above the cooler deeper waters. In most parts of the ocean, there is little exchange between these layers. Organisms that die in aquatic environments generally sink to the bottom, where they decompose, which releases nutrients in the form of nitrogen (N), phosphorus (P) and potassium (K). These nutrients are necessary for plant growth, but in the tropics, they do not directly return to the surface.

Plants form the base of the food chain, and need sunlight and nutrients to grow. In the ocean, these plants are mainly microscopic phytoplankton which drift in the water column. They need sunlight for photosynthesis, which powers carbon fixation, so they are found only relatively near the surface. But they also need nutrients. Phytoplankton rapidly use nutrients in the surface waters, and in the tropics, these nutrients are not usually replaced because of the thermocline.

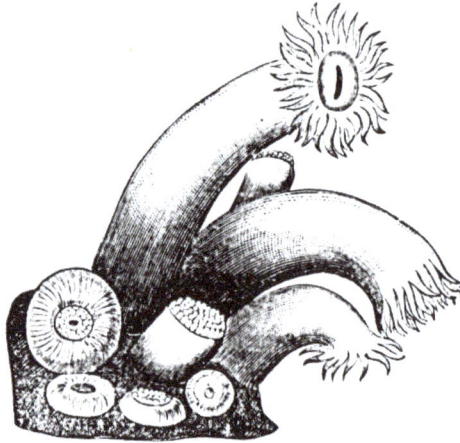

Coral polyps

Explanations

Around coral reefs, lagoons fill in with material eroded from the reef and the island. They become havens for marine life, providing protection from waves and storms.

Most importantly, reefs recycle nutrients, which happens much less in the open ocean. In coral reefs and lagoons, producers include phytoplankton, as well as seaweed and coralline algae, especially small types called turf algae, which pass nutrients to corals. The phytoplankton are eaten by fish and crustaceans, who also pass nutrients along the food web. Recycling ensures fewer nutrients are needed overall to support the community.

Coral reefs support many symbiotic relationships. In particular, zooxanthellae provide energy to coral in the form of glucose, glycerol, and amino acids. Zooxanthellae can provide up to 90% of a coral's energy requirements. In return, as an example of mutualism, the corals shelter the zooxanthellae, averaging one million for every cubic centimeter of coral, and provide a constant supply of the carbon dioxide they need for photosynthesis.

The color of corals depends on the combination of brown shades provided by their zooxanthellae and pigmented proteins (reds, blues, greens, etc.) produced by the corals themselves.

Corals also absorb nutrients, including inorganic nitrogen and phosphorus, directly from water. Many corals extend their tentacles at night to catch zooplankton that brush them when the water is agitated. Zooplankton provide the polyp with nitrogen, and the polyp shares some of the nitrogen with the zooxanthellae, which also require this element. The varying pigments in different species of zooxanthellae give them an overall brown or golden-brown appearance, and give brown corals their colors. Other pigments such as reds, blues, greens, etc. come from colored proteins made by the coral animals. Coral which loses a large fraction of its zooxanthellae becomes white (or sometimes pastel shades in corals that are richly pigmented with their own colorful proteins) and is said to be bleached, a condition which, unless corrected, can kill the coral.

Sponges are another key: they live in crevices in the coral reefs. They are efficient filter feeders, and in the Red Sea they consume about 60% of the phytoplankton that drifts by. The sponges eventually excrete nutrients in a form the corals can use.

Most coral polyps are nocturnal feeders. Here, in the dark, polyps have extended their tentacles to feed on zooplankton.

The roughness of coral surfaces is the key to coral survival in agitated waters. Normally, a boundary layer of still water surrounds a submerged object, which acts as a barrier. Waves breaking on the extremely rough edges of corals disrupt the boundary layer, allowing the corals access to passing nutrients. Turbulent water thereby promotes reef growth and branching. Without the nutritional gains brought by rough coral surfaces, even the most effective recycling would leave corals wanting in nutrients.

Studies have shown that deep nutrient-rich water entering coral reefs through isolated events may have significant effects on temperature and nutrient systems. This water movement disrupts the relatively stable thermocline that usually exists between warm shallow water to deeper colder water. Leichter et al. (2006) found that temperature regimes on coral reefs in the Bahamas and Florida were highly variable with temporal scales of minutes to seasons and spatial scales across depths.

Water can be moved through coral reefs in various ways, including current rings, surface waves, internal waves and tidal changes. Movement is generally created by tides and wind. As tides interact with varying bathymetry and wind mixes with surface water, internal waves are created. An internal wave is a gravity wave that moves along density stratification within the ocean. When a water parcel encounters a different density it will oscillate and create internal waves. While internal waves generally have a lower frequency than surface waves, they often form as a single wave that breaks into multiple waves as it hits a slope and moves upward. This vertical break up of internal waves causes significant diapycnal mixing and turbulence. Internal waves can act as nutrient pumps, bringing plankton and cool nutrient-rich water up to the surface.

The irregular structure characteristic of coral reef bathymetry may enhance mixing and produce pockets of cooler water and variable nutrient content. Arrival of cool, nutrient-rich water from depths due to internal waves and tidal bores has been linked to growth rates of suspension feeders and benthic algae as well as plankton and larval organisms. Leichter et al. proposed that Codium isthmocladum react to deep water nutrient sources due to their tissues having different concentrations of nutrients dependent upon depth. Wolanski and Hamner noted aggregations of eggs, larval organisms and plankton on reefs in response to deep water intrusions. Similarly, as internal waves and bores move vertically, surface-dwelling larval organisms are carried toward the shore. This has significant biological importance to cascading effects of food chains in coral reef ecosystems and may provide yet another key to unlocking "Darwin's Paradox".

Cyanobacteria provide soluble nitrates for the reef via nitrogen fixation.

Coral reefs also often depend on surrounding habitats, such as seagrass meadows and mangrove forests, for nutrients. Seagrass and mangroves supply dead plants and animals which are rich in nitrogen and also serve to feed fish and animals from the reef by supplying wood and vegetation. Reefs, in turn, protect mangroves and seagrass from waves and produce sediment in which the mangroves and seagrass can root.

Biodiversity

Tube sponges attracting cardinal fishes, glassfishes and wrasses

Over 4,000 species of fish inhabit coral reefs.

Organisms can cover every square inch of a coral reef.

Coral reefs form some of the world's most productive ecosystems, providing complex and varied marine habitats that support a wide range of other organisms.Fringing reefs just below low tide level have a mutually beneficial relationship with mangrove forests at high tide level and sea grass meadows in between: the reefs protect the mangroves and seagrass from strong currents and waves that would damage them or erode the sediments in which they are rooted, while the mangroves and sea grass protect the coral from large influxes of silt, fresh water and pollutants. This level of variety in the environment benefits many coral reef animals, which, for example, may feed in the sea grass and use the reefs for protection or breeding.

Reefs are home to a large variety of animals, including fish, seabirds, sponges, cnidarians (which includes some types of corals and jellyfish), worms, crustaceans (including shrimp, cleaner shrimp, spiny lobsters and crabs), mollusks (including cephalopods), echinoderms (including starfish, sea urchins and sea cucumbers), sea squirts, sea turtles and sea snakes. Aside from humans, mammals are rare on coral reefs, with visiting cetaceans such as dolphins being the main exception. A few of these varied species feed directly on corals, while others graze on algae on the reef. Reef biomass is positively related to species diversity.

The same hideouts in a reef may be regularly inhabited by different species at different times of day. Nighttime predators such as cardinalfish and squirrelfish hide during the day, while damselfish, surgeonfish, triggerfish, wrasses and parrotfish hide from eels and sharks.

Algae

Reefs are chronically at risk of algal encroachment. Overfishing and excess nutrient supply from onshore can enable algae to outcompete and kill the coral. Increased nutrient levels can be a result of sewage or chemical fertilizer runoff from nearby coastal developments. Runoff can carry nitrogen and phosphorus which promote excess algae growth. Algae can sometimes out-compete the coral for space. The algae can then smother the coral by decreasing the oxygen supply available to the reef. Decreased oxygen levels can slow down coral's calcification rates weakening the coral and leaving it more susceptible to disease and degradation. In surveys done around largely uninhabited US Pacific islands, algae inhabit a large percentage of surveyed coral locations. The algal population consists of turf algae, coralline algae, and macro algae.

Sponges

Sponges are essential for the functioning of the coral reef's ecosystem. Algae and corals in coral reefs produce organic material. This is filtered through sponges which convert this organic material into small particles which in turn are absorbed by algae and corals.

Fish

Over 4,000 species of fish inhabit coral reefs. The reasons for this diversity remain unclear. Hypotheses include the "lottery", in which the first (lucky winner) recruit to a territory is typically able to defend it against latecomers, "competition", in which adults compete for territory, and less-competitive species must be able to survive in poorer habitat, and "predation", in which population size is a function of postsettlement piscivore mortality. Healthy reefs can produce up to 35 tons of fish per square kilometer each year, but damaged reefs produce much less.

Invertebrates

Sea urchins, Dotidae and sea slugs eat seaweed. Some species of sea urchins, such as *Diadema antillarum*, can play a pivotal part in preventing algae from overrunning reefs. Nudibranchia and sea anemones eat sponges.

A number of invertebrates, collectively called "cryptofauna," inhabit the coral skeletal substrate itself, either boring into the skeletons (through the process of bioerosion) or living in pre-existing voids and crevices. Those animals boring into the rock include sponges, bivalve mollusks, and sipunculans. Those settling on the reef include many other species, particularly crustaceans and polychaete worms.

Seabirds

Coral reef systems provide important habitats for seabird species, some endangered. For example, Midway Atoll in Hawaii supports nearly three million seabirds, including two-thirds (1.5 million) of the global population of Laysan albatross, and one-third of the global population of black-footed albatross. Each seabird species has specific sites on the atoll where they nest. Altogether, 17 species of seabirds live on Midway. The short-tailed albatross is the rarest, with fewer than 2,200 surviving after excessive feather hunting in the late 19th century.

Other

Sea snakes feed exclusively on fish and their eggs. Marine birds, such as herons, gannets, pelicans and boobies, feed on reef fish. Some land-based reptiles intermittently associate with reefs, such as monitor lizards, the marine crocodile and semiaquatic snakes, such as *Laticauda colubrina*. Sea turtles, particularly hawksbill sea turtles, feed on sponges.

Whitetip reef shark

Green turtle

Giant clam

Importance

Coral reefs deliver ecosystem services to tourism, fisheries and coastline protection. The global economic value of coral reefs has been estimated to be between US $29.8 billion and $375 billion per year. Coral reefs protect shorelines by absorbing wave energy, and many small islands would not exist without their reefs to protect them. According to the environmental group World Wide Fund for Nature, the economic cost over a 25-year period of destroying one kilometer of coral reef is somewhere between $137,000 and $1,200,000. About six million tons of fish are taken each year from coral reefs. Well-managed coral reefs have an annual yield of 15 tons of seafood on average per square kilometer. Southeast Asia's coral reef fisheries alone yield about $2.4 billion annually from seafood.

To improve the management of coastal coral reefs, another environmental group, the World Resources Institute (WRI) developed and published tools for calculating the value of coral reef-related tourism, shoreline protection and fisheries, partnering with five Caribbean countries. As of April 2011, published working papers covered St. Lucia, Tobago, Belize, and the Dominican Republic, with a paper for Jamaica in preparation. The WRI was also "making sure that the study results support improved coastal policies and management planning". The Belize study estimated the value of reef and mangrove services at $395–559 million annually.

Bermuda's coral reefs provide economic benefits to the Island worth on average $722 million per year, based on six key ecosystem services, according to Sarkis *et al* (2010).

Threats

Island with fringing reef off Yap, Micronesia

Coral reefs are dying around the world. In particular, coral mining, agricultural and urban runoff, pollution (organic and inorganic), overfishing, blast fishing, disease, and the digging of canals and access into islands and bays are localized threats to coral ecosystems. Broader threats are sea temperature rise, sea level rise and pH changes from ocean acidification, all associated with greenhouse gas emissions. A 2014 study lists factors such as population explosion along the coast lines, overfishing, the pollution of coastal areas, global warming and invasive species among the main reasons that have put reefs in danger of extinction.

A study released in April 2013 has shown that air pollution can also stunt the growth of coral reefs; researchers from Australia, Panama and the UK used coral records (between 1880 and 2000) from the western Caribbean to show the threat of factors such as coal-burning coal and volcanic eruptions. Pollutants, such as Tributyltin, a biocide released into water from in anti-fouling paint can be toxic to corals.

In 2011, researchers suggested that "extant marine invertebrates face the same synergistic effects of multiple stressors" that occurred during the end-Permian extinction, and that genera "with poorly buffered respiratory physiology and calcareous shells", such as corals, were particularly vulnerable.

Rock coral on seamounts across the ocean are under fire from bottom trawling. Reportedly up to 50% of the catch is rock coral, and the practice transforms coral structures to rubble. With it taking years to regrow, these coral communities are disappearing faster than they can sustain themselves.

Another cause for the death of coral reefs is bioerosion. Various fishes graze corals, dead or alive and change the morphology of coral reefs making them more susceptible to other physical and chemical threats. It has been generally observed that only the algae growing on dead corals is eaten and the live ones are not. However, this act still destroys the top layer of coral substrate and makes it harder for the reefs to sustain.

In El Niño-year 2010, preliminary reports show global coral bleaching reached its worst level since another El Niño year, 1998, when 16% of the world's reefs died as a result of increased water temperature. In Indonesia's Aceh province, surveys showed some 80% of bleached corals died. Scientists do not yet understand the long-term impacts of coral bleaching, but they do know that bleaching leaves corals vulnerable to disease, stunts their growth, and affects their reproduction, while severe bleaching kills them. In July, Malaysia closed several dive sites where virtually all the corals were damaged by bleaching.

To find answers for these problems, researchers study the various factors that impact reefs. The list includes the ocean's role as a carbon dioxide sink, atmospheric changes, ultraviolet light, ocean acidification, viruses, impacts of dust storms carrying agents to far-flung reefs, pollutants, algal blooms and others. Reefs are threatened well beyond coastal areas. Coral reefs with one type of zooxanthellae are more prone to bleaching than are reefs with another, more hardy, species.

General estimates show approximately 10% of the world's coral reefs are dead. About 60% of the world's reefs are at risk due to destructive, human-related activities. The threat to the health of reefs is particularly high in Southeast Asia, where 95% of reefs are at risk from local threats. By the 2030s, 90% of reefs are expected to be at risk from both human activities and climate change; by 2050, all coral reefs will be in danger.

Current research is showing that ecotourism in the Great Barrier Reef is contributing to coral disease, and that chemicals in sunscreens may contribute to the impact of viruses on zooxanthellae.

Protection

A diversity of corals

Marine protected areas (MPAs) have become increasingly prominent for reef management. MPAs promote responsible fishery management and habitat protection. Much like national parks and wildlife refuges, and to varying degrees, MPAs restrict potentially damaging activities. MPAs encompass both social and biological objectives, including reef restoration, aesthetics, biodiversity, and economic benefits. However, there are very few MPAs that have actually made a substantial difference. Research in Indonesia, Philippines and Papua New Guinea shows that there is no significant difference between an MPA site and an unprotected site. Conflicts surrounding MPAs involve lack of participation, clashing views of the government and fisheries, effectiveness of the area, and funding. In some situations, as in the Phoenix Islands Protected Area, MPAs can also provide revenue, potentially equal to the income they would have generated without controls, as Kiribati did for its Phoenix Islands.

According to the *Caribbean Coral Reefs - Status Report 1970-2012* made by the IUCN. States that; stopping overfishing especially key fishes to coral reef like parrotfish, coastal zone management which reduce human pressure on reef, (for example restricting the coastal settlement, development and tourism in coastal reef) and controlling pollution specially sewage wastage, may not only reduce coral declining but also reverse it and may let to coral reef more adaptable to changes relates to climate and acidification. The report shows that healthier reef in the Caribbean are those with large population of parrotfish in countries which protect these key fishes and sea urchins, banning fish trap and Spearfishing creating "resilient reefs".

To help combat ocean acidification, some laws are in place to reduce greenhouse gases such as carbon dioxide. The Clean Water Act puts pressure on state government agencies to monitor and limit runoff of pollutants that can cause ocean acidification. Stormwater surge preventions are also in place, as well as coastal buffers between agricultural land and the coastline. This act also ensures that delicate watershed ecosystems are intact, such as wetlands. The Clean Water Act is funded by the federal government, and is monitored by various watershed groups. Many land use laws aim to reduce CO_2 emissions by limiting deforestation. Deforestation causes erosion, which releases a large amount of carbon stored in the soil, which then flows into the ocean, contributing to ocean acidification. Incentives are used to reduce miles traveled by vehicles, which reduces the carbon emissions into the atmosphere, thereby reducing the amount of dissolved CO_2 in the ocean. State and federal governments also control coastal erosion, which releases stored carbon in the soil into the ocean, increasing ocean acidification. High-end satellite technology is increasingly being employed to monitor coral reef conditions.

Biosphere reserve, marine park, national monument and world heritage status can protect reefs. For example, Belize's barrier reef, Chagos archipelago, Sian Ka'an, the Galapagos islands, Great Barrier Reef, Henderson Island, Palau and Papahānaumokuākea Marine National Monument are world heritage sites.

In Australia, the Great Barrier Reef is protected by the Great Barrier Reef Marine Park Authority, and is the subject of much legislation, including a biodiversity action plan. They have compiled a Coral Reef Resilience Action Plan. This detailed action plan consists of numerous adaptive management strategies, including reducing our carbon footprint, which would ultimately reduce the amount of ocean acidification in the oceans surrounding the Great Barrier Reef. An extensive public awareness plan is also in place to provide education on the "rainforests of the sea" and how people can reduce carbon emissions, thereby reducing ocean acidification.

Inhabitants of Ahus Island, Manus Province, Papua New Guinea, have followed a generations-old practice of restricting fishing in six areas of their reef lagoon. Their cultural traditions allow line fishing, but no net or spear fishing. The result is both the biomass and individual fish sizes are significantly larger than in places where fishing is unrestricted.

Restoration

Coral fragments growing on nontoxic concrete

Coral aquaculture, also known as coral farming or coral gardening, is showing promise as a potentially effective tool for restoring coral reefs, which have been declining around the world. The process bypasses the early growth stages of corals when they are most at risk of dying. Coral seeds are grown in nurseries, then replanted on the reef. Coral is farmed by coral farmers who live locally to the reefs and farm for reef conservation or for income.

Efforts to expand the size and number of coral reefs generally involve supplying substrate to allow more corals to find a home. Substrate materials include discarded vehicle tires, scuttled ships, subway cars, and formed concrete, such as reef balls. Reefs also grow unaided on marine structures such as oil rigs. In large restoration projects, propagated hermatypic coral on substrate can be secured with metal pins, superglue or milliput. Needle and thread can also attach A-hermatype coral to substrate.

A substrate for growing corals referred to as Biorock is produced by running low voltage electrical currents through seawater to crystallize dissolved minerals onto steel structures. The resultant white carbonate (aragonite) is the same mineral that makes up natural coral reefs. Corals rapidly colonize and grow at accelerated rates on these coated structures. The electrical currents also accelerate formation and growth of both chemical limestone rock and the skeletons of corals and other shell-bearing organisms. The vicinity of the anode and cathode provides a high-pH environment which inhibits the growth of competitive filamentous and fleshy algae. The increased growth rates fully depend on the accretion activity.

During accretion, the settled corals display an increased growth rate, size and density, but after the process is complete, growth rate and density return to levels comparable to natural growth, and are about the same size or slightly smaller.

One case study with coral reef restoration was conducted on the island of Oahu in Hawaii. The University of Hawaii has come up with a Coral Reef Assessment and Monitoring Program to help relocate and restore coral reefs in Hawaii. A boat channel on the island of Oahu to the Hawaii Institute of Marine Biology was overcrowded with coral reefs. Also, many areas of coral reef patches in the channel had been damaged from past dredging in the channel. Dredging covers the existing corals with sand, and their larvae cannot build and thrive on sand; they can only build on to existing reefs. Because of this, the University of Hawaii decided to relocate some of the coral reef to a different transplant site. They transplanted them with the help of the United States Army divers, to a relocation site relatively close to the channel. They observed very little, if any, damage occurred to any of the colonies while they were being transported, and no mortality of coral reefs has been observed on the new transplant site, but they will be continuing to monitor the new transplant site to see how potential environmental impacts (i.e. ocean acidification) will harm the overall reef mortality rate. While trying to attach the coral to the new transplant site, they found the coral placed on hard rock is growing considerably well, and coral was even growing on the wires that attached the transplant corals to the transplant site. This gives new hope to future research on coral reef transplant sites. As a result of this coral restoration project, no environmental effects were seen from the transplantation process, no recreational activities were decreased, and no scenic areas were affected by the project. This is a great example that coral transplantation and restoration can work and thrive under the right conditions, which means there may be hope for other damaged coral reefs.

Another possibility for coral restoration is gene therapy. Through infecting coral with genetically modified bacteria, it may be possible to grow corals that are more resistant to climate change and other threats.

Reefs in the Past

Ancient coral reefs

Throughout Earth history, from a few thousand years after hard skeletons were developed by marine organisms, there were almost always reefs. The times of maximum development were in the Middle Cambrian (513–501 Ma), Devonian (416–359 Ma) and Carboniferous (359–299 Ma), owing to order Rugosa extinct corals, and Late Cretaceous (100–66 Ma) and all Neogene (23 Ma–present), owing to order Scleractinia corals.

Not all reefs in the past were formed by corals: those in the Early Cambrian (542–513 Ma) resulted from calcareous algae and archaeocyathids (small animals with conical shape, probably related to sponges) and in the Late Cretaceous (100–66 Ma), when there also existed reefs formed by a group of bivalves called rudists; one of the valves formed the main conical structure and the other, much smaller valve acted as a cap.

Measurements of the oxygen isotopic composition of the aragonitic skeleton of coral reefs, such as Porites, can indicate changes in the sea surface temperature and sea surface salinity conditions of the ocean during the growth of the coral. This technique is often used by climate scientists to infer the paleoclimate of a region.

Deep Sea

The deep sea or deep layer is the lowest layer in the ocean, existing below the thermocline and above the seabed, at a depth of 1000 fathoms (1800 m) or more. Little or no light penetrates this part of the ocean, and most of the organisms that live there rely for subsistence on falling organic matter produced in the photic zone. For this reason, scientists once assumed that life would be sparse in the deep ocean, but virtually every probe has revealed that, on the contrary, life is abundant in the deep ocean.

From the time of Pliny until the late nineteenth century...humans believed there was no life in the deep. It took a historic expedition in the ship *Challenger* between 1872 and 1876 to prove Pliny wrong; its deep-sea dredges and trawls brought up living things from all depths that could be reached. Yet even in the twentieth century scientists continued to imagine that life at great depth was insubstantial, or somehow inconsequential. The eternal dark, the almost inconceivable pressure, and the extreme cold that exist below one thousand meters were, they thought, so forbidding as to have all but extinguished life. The reverse is in fact true....(Below 200 meters) lies the largest habitat on earth.

In 1960, the Bathyscaphe *Trieste* descended to the bottom of the Mariana Trench near Guam, at 10,911 m (35,797 ft; 6.780 mi), the deepest known spot in any ocean. If Mount Everest (8,848 metres) were submerged there, its peak would be more than a mile beneath the surface. The *Trieste* was retired, and for a while the Japanese remote-operated vehicle (ROV) Kaikō was the only vessel capable of reaching this depth. It was lost at sea in 2003. In May and June 2009, the hybrid-ROV (HROV) *Nereus* returned to the Challenger Deep for a series of three dives to depths exceeding 10900 meters.

It has been suggested that more is known about the Moon than the deepest parts of the ocean. Little was known about the extent of life on the deep ocean floor until the discovery of thriving colonies of shrimps and other organisms around hydrothermal vents in the late 1970s. Before the discovery of the undersea vents, it had been accepted that almost all life on earth obtained its energy (one way or another) from the sun. The new discoveries revealed groups of creatures that obtained nutrients and energy directly from thermal sources and chemical reactions associated with changes to mineral deposits. These organisms thrive in completely lightless and anaerobic environments in highly saline water that may reach 300 °F (150 °C), drawing their sustenance from hydrogen sulfide, which is highly toxic to almost all terrestrial life. The revolutionary discovery that life can exist under these extreme conditions changed opinions about the chances of there being life elsewhere in the universe. Scientists now speculate that Europa, one of Jupiter's moons, may be able to support life beneath its icy surface, where there is evidence of a global ocean of liquid water.

Environmental Characteristics

Light

Natural light does not penetrate the deep ocean, with the exception of the upper parts of the mesopelagic. Since photosynthesis is not possible, plants cannot live in this zone. Since plants are the primary producers of almost all of earth's ecosystems, life in this area of the ocean must depend on energy sources from elsewhere. Except for the areas close to the hydrothermal vents, this energy comes from organic material drifting down from the photic zone. The sinking organic material is composed of algal particulates, detritus, and other forms of biological waste, which is collectively referred to as marine snow.

Pressure

Because pressure in the ocean increases by about 1 atmosphere for every 10 meters of depth, the amount of pressure experienced by many marine organisms is extreme. Until recent years, the scientific community lacked detailed information about the effects of pressure on most deep sea organisms because the specimens encountered arrived at the surface dead or dying and weren't observable at the pressures at which they lived. With the advent of traps that incorporate a special pressure-maintaining chamber, undamaged larger metazoan animals have been retrieved from the deep sea in good condition.

Salinity

Salinity is remarkably constant throughout the deep sea, at about 35 parts per thousand. There are some minor differences in salinity, but none that is ecologically significant, except in the Mediterranean and Red Seas.

Temperature

The two areas of greatest and most rapid temperature change in the oceans are the transition zone between the surface waters and the deep waters, the thermocline, and the transition between the deep-sea floor and the hot water flows at the hydrothermal vents. Thermoclines vary in thickness from a few hundred meters to nearly a thousand meters. Below the thermocline, the water mass of the deep ocean is cold and far more homogeneous. Thermoclines are strongest in the tropics, where the temperature of the epipelagic zone is usually above 20 °C. From the base of the epipelagic, the temperature drops over several hundred meters to 5 or 6 °C at 1,000 meters. It continues to decrease to the bottom, but the rate is much slower. Below 3,000 to 4,000 m, the water is isothermal between 0 to 3 °C. The cold water stems from sinking heavy surface water in the polar regions.

At any given depth, the temperature is practically unvarying over long periods of time. There are no seasonal temperature changes, nor are there any annual changes. No other habitat on earth has such a constant temperature.

Hydrothermal vents are the direct contrast with constant temperature. In these systems, the temperature of the water as it emerges from the "black smoker" chimneys may be as high as 400 °C (it is kept from boiling by the high hydrostatic pressure) while within a few meters it may be back down to 2 - 4 °C.

Biology

Regions below the epipelagic are divided into further zones, beginning with the *mesopelagic* which spans from 200 to 1000 meters below sea level, where a little light penetrates while still being insufficient for primary production. Below this zone the deep sea begins, consisting of the aphotic *bathypelagic*, *abyssopelagic* and *hadopelagic*. Food consists of falling organic matter known as 'marine snow' and carcasses derived from the productive zone above, and is scarce both in terms of spatial and temporal distribution.

Instead of relying on gas for their buoyancy, many species have jelly-like flesh consisting mostly of glycosaminoglycans, which has very low density. It is also common among deep water squid to combine the gelatinous tissue with a flotation chamber filled with a coelomic fluid made up of the metabolic waste product ammonium chloride, which is lighter than the surrounding water.

The midwater fish have special adaptations to cope with these conditions—they are small, usually being under 25 centimetres (10 in); they have slow metabolisms and unspecialized diets, preferring to sit and wait for food rather than waste energy searching for it. They have elongated bodies with weak, watery muscles and skeletal structures. They often have extendable, hinged jaws with recurved teeth. Because of the sparse distribution and lack of light, finding a partner with which to breed is difficult, and many organisms are hermaphroditic.

Because light is so scarce, fish often have larger than normal, tubular eyes with only rod cells. Their upward field of vision allows them to seek out the silhouette of possible prey. Prey fish however also have adaptations to cope with predation. These adaptations are mainly concerned with reduction of silhouettes, a form of camouflage. The two main methods by which this is achieved are reduction in the area of their shadow by lateral compression of the body, and counter illumination via bioluminescence. This is achieved by production of light from ventral photophores, which tend to produce such light intensity to render the underside of the fish of similar appearance to the background light. For more sensitive vision in low light, some fish have a retroreflector behind the retina. Flashlight fish have this plus photophores, which combination they use to detect eyeshine in other fish (see *tapetum lucidum*).

Organisms in the deep sea are almost entirely reliant upon sinking living and dead organic matter which falls at approximately 100 meters per day. In addition, only about 1-3% of the production from the surface reaches the sea bed mostly in the form of marine snow. Larger food falls, such as whale carcasses, also occur and studies have shown that these may happen more often than currently believed. There are many scavengers that feed primarily or entirely upon large food falls and the distance between whale carcasses is estimated to only be 8 kilometers. In addition, there are a number of filter feeders that feed upon organic particles using tentacles, such as *Freyella elegans*.

Marine bacteriophages play an important role in cycling nutrients in deep sea sediments. They are extremely abundant (between 5×10^{12} and 1×10^{13} phages per square meter) in sediments around the world.

Chemosynthesis

There are a number of species that do not primarily rely upon dissolved organic matter for their food and these are found at hydrothermal vents. One example is the symbiotic relationship between the tube worm *Riftia* and chemosynthetic bacteria. It is this chemosynthesis that supports the complex communities that can be found around hydrothermal vents. These complex communities are one of the few ecosystems on the planet that do not rely upon sunlight for their supply of energy.

Exploration

Describing the operation and use of an autonomous lander (RV Kaharoa) in deep-sea research; the fish seen are the abyssal grenadier (*Coryphaenoides armatus*).

The deep sea is an environment completely unfriendly to mankind; it represents one of the least explored areas on Earth. Pressures even in the mesopelagic become too great for traditional exploration methods, demanding alternative approaches for deep sea research. Baited camera stations, small manned submersibles and ROVs (remotely operated vehicles) are three methods utilized to explore the ocean's depths. Because of the difficulty and cost of exploring this zone, current knowledge is limited. Pressure increases at approximately one atmosphere for every 10 meters meaning that some areas of the deep sea can reach pressures of above 1,000 atmospheres. This not only makes great depths very difficult to reach without mechanical aids, but also provides a significant difficulty when attempting to study any organisms that may live in these areas as their cell chemistry will be adapted to such vast pressures.

Benthos

Benthos is the community of organisms that live on, in, or near the seabed, also known as the benthic zone. This community lives in or near marine sedimentary environments, from tidal pools along the foreshore, out to the continental shelf, and then down to the abyssal depths.

Many organisms adapted to deep-water pressure cannot survive in the upper parts of the water column. The pressure difference can be very significant (approximately one atmosphere for each 10 metres of water depth).

Because light is absorbed before it can reach deep ocean-water, the energy source for deep benthic ecosystems is often organic matter from higher up in the water column that drifts down to the depths. This dead and decaying matter sustains the benthic food chain; most organisms in the benthic zone are scavengers or detritivores.

The term *benthos*, coined by Haeckel in 1891, comes from the Greek noun βένθος "depth of the sea". *Benthos* is also used in freshwater biology to refer to organisms at the bottom of freshwater bodies of water, such as lakes, rivers, and streams.

Food Sources

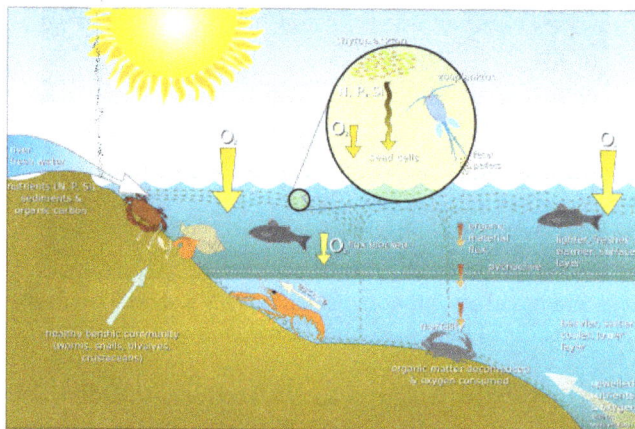

Effect of eutrophication on marine benthic life

The main food sources for the benthos are algae and organic runoff from land. The depth of water, temperature and salinity, and type of local substrate all affect what benthos is present. In coastal waters and other places where light reaches the bottom, benthic photosynthesizing diatoms can proliferate. Filter feeders, such as sponges and bivalves, dominate hard, sandy bottoms. Deposit feeders, such as polychaetes, populate softer bottoms. Fish, such as dragonets, as well as sea stars, snails, cephalopods, and crustaceans are important predators and scavengers.

Benthic organisms, such as sea stars, oysters, clams, sea cucumbers, brittle stars and sea anemones, play an important role as a food source for fish, such as the California sheephead, and humans.

By size

Macrobenthos

Macrobenthos comprises the larger, more visible, benthic organisms that are greater than 1 mm in size. Some examples are polychaete worms, bivalves, echinoderms, sea anemones, corals, sponges, sea squirts, turbellarians and larger crustaceans such as crabs, lobsters and cumaceans.

They are easily visible to the naked eye with the lower range of body size at 0.5 mm but usually larger than 3 mm. In the coastal water ecosystem, they include several species of organisms from different taxa including Porifera, Annelids, Coelenterates, Mollusks, Crustaceans, Arthropods etc.

Seagrass growing off the coast

Echinoderms

A sea squirt being used as a substrate for a nudibranch's spiral egg

Microphotograph of typical macrobenthic animals, (from top to bottom) including amphipods, a poly-chaete worm, a snail, and a chironomous midge larvae

Meiobenthos

Meiobenthos comprises tiny benthic organisms that are less than 1 mm but greater than 0.1 mm in size. Some examples are nematodes, foraminiferans, water bears, gastrotriches and smaller crustaceans such as copepods and ostracodes

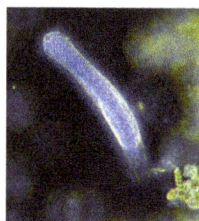

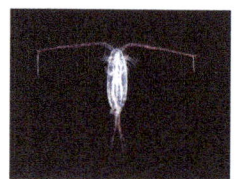

Live foraminifera *Ammonia tepida* (Rotaliida)

Water bear *Hypsibius dujardini*

Gastrotrich

Copepod

Microbenthos

Microbenthos comprises microscopic benthic organisms that are less than 0.1 mm in size. Some examples are bacteria, diatoms, ciliates, amoeba, flagellates.

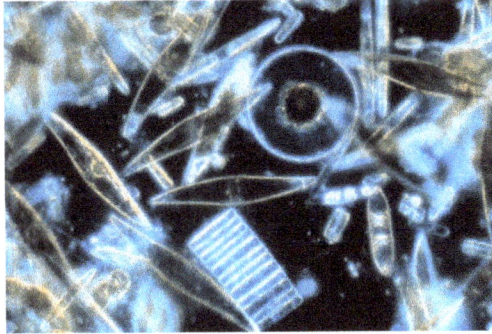
Marine diatoms

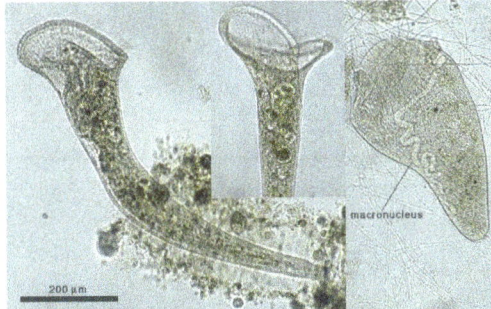
Ciliate *stentor roeseli*

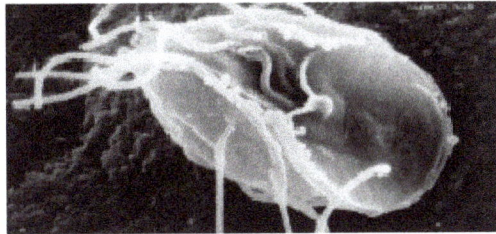
Flagellate

By Type

Zoobenthos

Zoobenthos comprises the animals belonging to the benthos.

Phytobenthos

Phytobenthos comprises the plants belonging to the benthos, mainly benthic diatoms and macroalgae (seaweed).

By location

Endobenthos

Endobenthos lives buried, or burrowing in the sediment, often in the oxygenated top layer, e.g., a sea pen or a sand dollar.

Epibenthos

Epibenthos lives on top of the sediments, e.g., like a sea cucumber or a sea snail crawling about.

Hyperbenthos

Hyperbenthos lives just above the sediment, e.g., a rock cod.

References

- Kennish, M. J. (1986). Ecology of Estuaries. Volume I: Physical and Chemical Aspects. Boca Raton, FL: CRC Press. ISBN 0-8493-5892-2.

- McLusky, D. S.; Elliott, M. (2004). The Estuarine Ecosystem: Ecology, Threats and Management. New York: Oxford University Press. ISBN 0-19-852508-7.

- Roach, Peter (2011), Cambridge English Pronouncing Dictionary (18th ed.), Cambridge: Cambridge University Press, ISBN 9780521152532

- Davis, Richard A., Jr. (1994). The Evolving Coast. New York: Scientific American Library. pp. 101, 107. ISBN 9780716750420.

- Allaby, Michael, ed. (1990). Oxford Dictionary of Earth Sciences. Oxford: Oxford University Press. ISBN 978-0-19-921194-4.

- Nybakken, James W., ed. (2003). Interdisciplinary Encyclopedia of Marine Sciences. 2 G-O. Danbury, Connecticutt: Grolier Academic Reference. pp. 189–90. ISBN 0-7172-5946-3.

- Bird, Eric C. F. (2010). Encyclopedia of the World's Coastal Landforms, Volume 1. Dordrecht: Springer. p. 485. ISBN 978-1-4020-8638-0.

- Xavier Romero-Frias, The Maldive Islanders, A Study of the Popular Culture of an Ancient Ocean Kingdom. Barcelona 1999, ISBN 84-7254-801-5

- Millennium Ecosystem Assessment (2005) Ecosystems and Human Well-being: Synthesis (p.2) Island Press, Washington, DC. World Resources Institute ISBN 1-59726-040-1

- Moyle, Peter B.; Joseph J. Cech (2004). Fishes : an introduction to ichthyology (Fifth ed.). Upper Saddle River, N.J.: Pearson/Prentice Hall. p. 556. ISBN 978-0-13-100847-2.

- Marshall, Paul; Schuttenberg, Heidi (2006). A Reef Manager's Guide to Coral Bleaching. Townsville, Australia: Great Barrier Reef Marine Park Authority. ISBN 1-876945-40-0.

- Crossland CJ (1983) "Dissolved nutrients in coral reef waters In DJ Barnes (Ed) Perspectives on coral reefs, pages 56–68, Australian Institute of Marine Science. ISBN 9780642895851.

- Glynn, P.W. (1990). Dubinsky, Z., ed. Ecosystems of the World v. 25-Coral Reefs. New York, NY: Elsevier Science. ISBN 978-0-444-87392-7.

- Osborne, Patrick L. (2000). Tropical Ecosystem and Ecological Concepts. Cambridge: Cambridge University Press. p. 464. ISBN 0-521-64523-9.

- R. N. Gibson, Harold (CON) Barnes, R. J. A. Atkinson, Oceanography and Marine Biology, An Annual Review. 2007. Volume 41. Published by CRC Press, 2004 ISBN 0-415-25463-9, ISBN 978-0-415-25463-2

- Ross, D. A. (1995). Introduction to Oceanography. New York: Harper Collins College Publishers.

ISBN 978-0-673-46938-0.

- Jennings S, Kaiser MJ and Reynolds JD (2001) Marine fisheries ecology, Wiley-Blackwell, pp. 291–293. ISBN 978-0-632-05098-7.

- "O Valor da Opção de Preservação do Parque dos Manguezais em Recife-PE: Uma Utilização do Método de Opções Reais" (PDF) (in Portuguese). ANPEC. Retrieved 2015-06-04.

- Somiya, Kazuo. "Conservation of landscape and culture in southwestern islands of Japan" (PDF). Naha Nature Conservation Office, Ministry of the Environment. Retrieved 2015-08-19.

- "THE EFFECTS OF TERRESTRIAL RUNOFF OF SEDIMENTS, NUTRIENTS AND OTHER POLLUTANTS ON CORAL REEFS" (PDF). Retrieved 2015-12-05.

- Sarkis, Samia; van Beukering, Pieter J.H.; McKenzie, Emily (2010). "Total Economic Value of Bermuda's Coral Reefs. Valuation of ecosystem Services" (PDF). Retrieved May 29, 2015.

- Ewa Magiera; Sylvie Rockel (2 July 2014). "From despair to repair: Dramatic decline of Caribbean corals can be reversed". Retrieved 8 June 2015.

Conservation of Marine Ecosystems

The protection and conservation of marine ecosystems is termed as marine conservation. It aims to limit the harm caused by humans on the marine ecosystem as well as restoring damaged ecosystems. The aspects elucidated in this text are of vital importance, and provide a better understanding of marine ecosystems.

Marine Resources Conservation

Marine conservation, also known as marine resources conservation, is the protection and preservation of ecosystems in oceans and seas. Marine conservation focuses on limiting human-caused damage to marine ecosystems, and on restoring damaged marine ecosystems. Marine conservation also focuses on preserving vulnerable marine species.

Overview

Marine conservation is a response to biological issues such as extinction and habitat change. Marine conservation is the study of conserving physical and biological marine resources and ecosystem functions. This is a relatively new discipline. Marine conservationists rely on a combination of scientific principles derived from marine biology, oceanography, and fisheries science, as well as on human factors such as demand for marine resources and marine law, economics and policy in order to determine how to best protect and conserve marine species and ecosystems. Marine conservation can be seen as sub discipline of conservation biology.

Coral Reefs

Coral reefs are the epicenter for immense amounts of biodiversity, and are a key player in the survival of an entire ecosystem. They provide various marine animals with food, protection, and shelter which keep generations of species alive. Furthermore, coral reefs are an integral part of sustaining human life through serving as a food source (i.e. fish, mollusks, etc.) as well as a marine space for eco-tourism which provides economic benefits.

Unfortunately, because of human impact of coral reefs, these ecosystems are becoming increasingly degraded and in need of conservation. The biggest threats include "overfishing, destructive fishing practices, and sedimentation and pollution from land-based sources." This in conjunction with increased carbon in oceans, coral bleaching, and diseases, there are no pristine reefs anywhere in the world. In fact, up to 88% of coral reefs in Southeast Asia are now threatened, with 50% of those reefs at either "high" or "very high" risk of disappearing which directly effects biodiversity and survival of species dependent on coral.

This is especially harmful to island nations such as Samoa, Indonesia, and the Philippines because many people depend on the coral reef ecosystems to feed their families and to make a living. However, many fisherman are unable to catch as many fish as they used to, so they are increasingly using cyanide and dynamite in fishing, which further degrades the coral reef ecosystem. This perpetuation of bad habits simply leads to the further decline of coral reefs and therefore perpetuating the problem. One solution to stopping this cycle is to educate the local community about why conservation of marine spaces that include coral reefs is important. Once the local communities understand the personal stakes at risk then they will actually fight to preserve the reefs. Conserving coral reefs has many economic, social, and ecological benefits, not only for the people who live on these islands, but for people throughout the world as well.

Human Impact

The deterioration of coral reefs is mainly linked to human activities – 88% of coral reefs are threatened through various reasons as listed above, including excessive amounts of CO_2 (Carbon Dioxide) emissions. Oceans absorb approximately 1/3 of the CO_2 produced by humans, which has detrimental effects on the marine environment. The increasing levels of CO_2 in oceans change the seawater chemistry by decreasing the level of pH. This process is also known as acidification. Acidification negatively affects the carbonate buffering system and drops the carbonate saturation by 30%, which results in a decrease in reef calcification. Reductions in calcification have negative implications on calcifiers, such as corals and shellfish. Some examples include diminishing coral resilience from bleaching, decreasing organisms' ability to fight off predators, inhibiting their potential to compete for food, and altering behavior patterns. When the bottom of the food web declines tremendously due to acidification, the food web and the whole marine conservation effort is jeopardized. Although humans cause the greatest threat to our marine environment, humans also have the ability to create effective management plans that will be the key to successful marine conservation. Although the most widely known conservation tool is the MPA, one of the best marine conservation tools simply stems from smarter individualist choices we make in efforts to reduce CO_2 emissions on a daily basis.

Techniques

Strategies and techniques for marine conservation tend to combine theoretical disciplines, such as population biology, with practical conservation strategies, such as setting up protected areas, as with marine protected areas (MPAs) or Voluntary Marine Conservation Areas. Other techniques include developing sustainable fisheries and restoring the populations of endangered species through artificial means.

Another focus of conservationists is on curtailing human activities that are detrimental to either marine ecosystems or species through policy, techniques such as fishing quotas, like those set up by the Northwest Atlantic Fisheries Organization, or laws such as those listed below. Recognizing the economics involved in human use of marine ecosystems is key, as is education of the public about conservation issues. This includes educating tourists that come to an area that might not be familiar of certain rules and regulations regarding the marine habitat. One example of this is a project called Green Fins that uses the SCUBA diving industry to educate the public based in SE Asia. This project, implemented by UNEP, encourages scuba diving operators to educate the public they teach to dive about the importance of marine conservation and encourage them to dive in an environmentally friendly manner that does not damage coral reefs or associated marine ecosystems.

Technology and Halfway Technology

Marine conservation technologies are devices used to protect endangered and threatened marine organisms and/or habitat. Marine conservation technologies are innovative and revolutionary because they reduce bycatch, increase the survivorship and health of marine life and habitat, and benefit fishermen who depend on the resources for profit. Examples of technologies include marine protected areas (MPAs), turtle excluder devices (TEDs), Autonomous recording unit, pop-up satellite archival tag, and radio-frequency identification (RFID). Commercial practicality plays in important role in the success of marine conservation because it is necessary to cater to the needs of fishermen while also protecting marine life.

Pop-up satellite archival tag (PSAT or PAT) serve a vital role in marine conservation by providing marine biologists with an opportunity to study animals in their natural environments. They are used to track movements of (usually large, migratory) marine animals. A PSAT (also commonly referred to as a PAT tag) is an archival tag (or data logger) that is equipped with a means to transmit the collected data via satellite. Though the data are physically stored on the tag, its major advantage is that it does not have to be physically retrieved like an archival tag for the data to be available making it a viable, fishery independent tool for animal behavior studies. They have been used to track movements of ocean sunfish, marlin, blue sharks, bluefin tuna, swordfish and sea turtles. Location, depth, temperature, and body movement data are used to answer questions about migratory patterns, seasonal feeding movements, daily habits, and survival after catch and release, for examples.

Another example, Turtle excluder devices (TEDs) remove a major threat to turtles in their marine environment. Many sea turtles are accidentally captured, injured or killed by fishing. In response to this threat the National Oceanic and Atmospheric Administration (NOAA)worked with the shrimp trawling industry to create the TEDs devices. By working with the industry they insured the commercial viability of the devices. Basically, a TED is a series of bars that is placed at the top or bottom of a trawl net, fitting the bars into the "neck" of the shrimp trawl and acting as a filter to ensure that only small animals may pass through. The shrimp will be caught but larger animals such as marine turtles that become caught by the trawler will be rejected by the filter function of the bars.

Similarly, halfway technologies work to increase the population of marine organisms, however, it does so without behavioral changes and "addresses the symptoms but not the cause of the declines". Examples of halfway technologies would include hatcheries and fish ladders.

Laws and Treaties

International laws and treaties related to marine conservation include the 1966 Convention on Fishing and Conservation of Living Resources of the High Seas. United States laws related to marine conservation include the 1972 Marine Mammal Protection Act, as well as the 1972 Marine Protection, Research and Sanctuaries Act which established the National Marine Sanctuaries program.

In 2010, the Scottish Parliament enacted new legislation for the protection of marine life with the Marine (Scotland) Act 2010. The provisions in the Act include: Marine planning, Marine licensing, marine conservation, seal conservation, and enforcement.

Organizations and Education

The shore of the Pacific Ocean in San Francisco, California.

There are marine conservation organizations throughout the world that focus on funding conservation efforts, educating the public and stakeholders, and lobbying for conservation law and policy. Examples of these organizations are Oceana (non-profit group), the Marine Conservation Institute (United States), Blue Frontier Campaign

(United States), Sea Shepherd Conservation Society [international], Frontier (the Society for Environmental Exploration) (United Kingdom), Marine Conservation Society (United Kingdom), Community Centred Conservation (C3), The Reef-World Foundation (United Kingdom), Reef Watch (India), and Australian Marine Conservation Society. Zoox (United Kingdom) is an example of an organisation that provides both marine conservation training and professional career development to volunteers who are also working on marine conservation projects such as Green Fins.

On a regional level, PERSGA- the Regional Organization for the Conservation of the Environment of the Red Sea and the Gulf of Aden, is a regional entity serves as the secretariat for the Jeddah Convention-1982, one of the first regional marine agreements. PERSGA Member States are: Djibouti, Egypt, Jordan, Saudi Arabia, Somalia, Sudan and Yemen.

Extinct and Endangered Species

Marine Mammals

Baleen whales were predominantly hunted from 1600 through the mid 1900s and were nearing extinction when a global ban on commercial whaling was put into effect in 1896 by the IWC (International Whaling Convention). The Atlantic gray Whale, last sited in 1740, is now extinct due to European and Native American Whaling. Since the 1960s the global population of Monk seals has been rapidly declining. The Hawaiian and Mediterranean monk seals are considered to be one of the most endangered marine mammals on the planet according to the NOAA. The last siting of the Caribbean monk seal was in 1952, it has now been confirmed extinct by the NOAA. The Vaquita porpoise, discovered in 1958, has become the most endangered marine species. Over half the population has disappeared since 2012, leaving 100 left in 2014. The Vaquita frequently drowns in fishing nets, which are used illegally in marine protected areas off the Gulf of Mexico.

Sea Turtles

In 2004, The Marine Turtle Specialist Group (MTSG), from the International Union for Conservation of Nature (IUCN) ran a Green Turtle Assessment that determined Green Turtles were globally endangered. Population decline in ocean basins over the last 100–150 years is indicated through data collected by the MTSG that analyzes abundance and historical information on the species. The data collected by MTSG examined the global population of the Green Turtles at 32 nesting sites. This data determined that over the last 100–150 years there has been a 48-65 percent decrease in the amount of mature nesting females. The Kemp's ridley sea turtle population fell in 1947 when 33,000 nests, which accounted for 80 percent of the population, were collected and sold by villagers in Racho Nuevo, Mexico. In the early 1960s only 5,000 individuals were left and between 1978 and 1991 200 Kemp's ridley turtles nested annually. In

2015, the World Wildlife Fund (WWF) and National Geographic Magazine named the Kemp's ridley the most endangered sea turtle in the world with 1000 females nesting annually.

Fish

In 2014, the IUCN moved the Pacific bluefin tuna from "least concerned" to "vulnerable" on a scale that represents level of extinction risk. The Pacific bluefin tuna is targeted by the fishing industry mainly for its use in sushi. A stock assessment released in 2013 by the International Scientific Committee for Tuna and Tuna-Like Species in the North Pacific Ocean (ISC) shows that the Pacific bluefin tuna population dropped by 96 percent in the Pacific Ocean. According to the ISC assessment, 90 percent of the Pacific bluefin tuna caught are juveniles that have not reproduced. Between the years 2011 and 2014, the European eel, Japanese eel, and American eel were put on the IUCN red list of endangered species. In 2015, The Environmental Agency concluded that the number of European eels has declined by 95 percent since 1990. An Environmental Agency officer, Andy Don who has been researching eels for the past 20 years says, "There is no doubt that there is a crisis. People have been reporting catching a kilo of glass eels this year when they would expect to catch 40 kilos. We have got to do something."

Marine Plants

Johnson's seagrass, a food source for the endangered Green sea turtle, is the scarcest species in its genius. It reproduces asexually which limits it ability to populate and colonize habitats. Data on this species is limited but since the 1970s there has been a 50 percent decrease in abundance.

History of Marine Conservation

Modern Marine conservation first became globally recognized in the 1970s after World War II in an era known as the marine revolution. The United States legislation showed its support of Marine conservation by institutionalizing protected areas, and creating marine estuaries. In the mid-1970s the United States formed the International Union for Conservation of Nature, the IUCN. Through this program, different nations could communicate and make agreements surrounding the topic of Marine conservation. After the formation of the IUCN new independent organizations known as NGOs started to appear. These organizations were self-governed and had individual goals for Marine conservation. At the end of the 1970s undersea explorations equipped with new technology such as computers were undergone. During these explorations, fundamental principles of change were discovered in relation to marine ecosystems. Through this discovery, the interdependent nature of the ocean was revealed. This discovery led to a change in the approach of marine conservation efforts and a new emphasis was put on restoring systems within the environment along with protecting biodiversity.

Overabundance

Overabundance occurs when the population of a certain species cannot be controlled. A domination of a certain species can create an imbalance in an ecosystem, which can lead to the demise of other species and of the habitat. Overabundance occurs predominately in invasive species. Cargo ships introduce new species into different environments through releasing ballast water into an ecosystem. A tank of ballast water is estimated to contain around 3,000 non-native species. The San Francisco Bay is one of the places in the world that is the most impacted by foreign and invasive species. According to the Baykeeper organization, 97 percent of the organisms in the San Francisco Bay have been compromised by the 240 invasive species that have been brought into the ecosystem. Invasive species in the San Francisco Bay such as the Asian clam (Corbicula fluminea) have changed the food web of the ecosystem by depleting populations of native species such as plankton. The Asian clam clogs pipes and obstructs the flow of water in electrical generating facilities. Their presence in the San Francisco Bay has cost the United States an estimated one billion dollars in damages.

References

- Carleton, Ray G.; McCormick, Jerry (1 April 2009). Coastal-Marine Conservation: Science and Policy. John Wiley & Sons. ISBN 978-1-4443-1124-2.

- Aidan Bodeo-Lomicky (4 February 2015). The Vaquita: The Biology of an Endangered Porpoise. Createspace Independent Pub. ISBN 978-1-5077-5577-8.

- Johnson's Seagrass (Halophila Johnsonii) - Office of Protected Resources - NOAA Fisheries. 1 March 2013. 19 February 2015.

- Martin, Glen. "The Great Invaders / A New Ecosystem Is Evolving in San Francisco Bay. We Have No Idea What It Is, or Where It's Going" SFGate. 5 February 2006. 19 February 2015.

- Foster, A.M., P. Fuller, A. Benson, S. Constant, D. Raikow, J. Larson, and A. Fusaro. 2015. Corbicula fluminea USGS Nonindigenous Aquatic Species Database, Gainesville, FL. Revision Date: 26 June 2014

Aquatic Ecosystems: An Integrated Study

The ecosystem that is found in a body of water is known as aquatic ecosystem. The two kinds of aquatic ecosystems are marine and freshwater ecosystem. Aquatic ecosystems also include freshwater ecosystem, lake ecosystem and river ecosystem. This section provides the reader with an integrated understanding of aquatic ecosystems.

Aquatic Ecosystem

An aquatic ecosystem is an ecosystem in a body of water. Communities of organisms that are dependent on each other and on their environment live in aquatic ecosystems. The two main types of aquatic ecosystems are marine ecosystems and freshwater ecosystems.

Types

Marine

Marine ecosystems cover approximately 71% of the Earth's surface and contain approximately 97% of the planet's water. They generate 32% of the world's net primary production. They are distinguished from freshwater ecosystems by the presence of dissolved compounds, especially salts, in the water. Approximately 85% of the dissolved materials in seawater are sodium and chlorine. Seawater has an average salinity of 35 parts per thousand (ppt) of water. Actual salinity varies among different marine ecosystems.

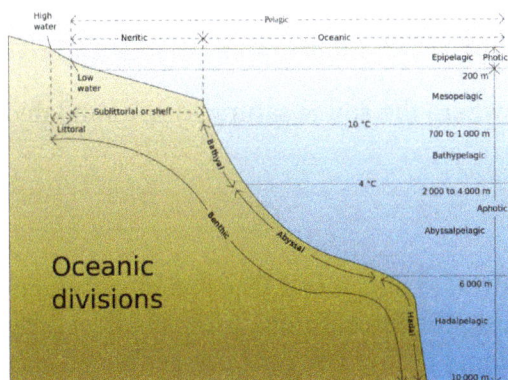

A classification of marine habitats.

Marine ecosystems can be divided into many zones depending upon water depth and shoreline features. The oceanic zone is the vast open part of the ocean where animals such as whales, sharks, and tuna live. The benthic zone consists of substrates below water where many invertebrates live. The intertidal zone is the area between high and low tides; in this figure it is termed the littoral zone. Other near-shore (neritic) zones can include estuaries, salt marshes, coral reefs, lagoons and mangrove swamps. In the deep water, hydrothermal vents may occur where chemosynthetic sulfur bacteria form the base of the food web.

Classes of organisms found in marine ecosystems include brown algae, dinoflagellates, corals, cephalopods, echinoderms, and sharks. Fishes caught in marine ecosystems are the biggest source of commercial foods obtained from wild populations.

Environmental problems concerning marine ecosystems include unsustainable exploitation of marine resources (for example overfishing of certain species), marine pollution, climate change, and building on coastal areas.

Freshwater

Freshwater ecosystem.

Freshwater ecosystems cover 0.80% of the Earth's surface and inhabit 0.009% of its total water. They generate nearly 3% of its net primary production. Freshwater ecosystems contain 41% of the world's known fish species.

There are three basic types of freshwater ecosystems:

- Lentic: slow moving water, including pools, ponds, and lakes.

- Lotic: faster moving water, for example streams and rivers.

- Wetlands: areas where the soil is saturated or inundated for at least part of the time.

Lentic

Lake ecosystems can be divided into zones. One common system divides lakes into three zones (see figure). The first, the littoral zone, is the shallow zone near the shore. This is where rooted wetland plants occur. The offshore is divided into two further zones, an open water zone and a deep water zone. In the open water zone (or photic zone) sunlight supports photosynthetic algae, and the species that feed upon them. In the deep water zone, sunlight is not available and the food web is based on detritus entering from the littoral and photic zones. Some systems use other names. The off shore areas may be called the pelagic zone, the photic zone may be called the limnetic zone and the aphotic zone may be called the profundal zone. Inland from the littoral zone one can also frequently identify a riparian zone which has plants still affected by the presence of the lake—this can include effects from windfalls, spring flooding, and winter ice damage. The production of the lake as a whole is the result of production from plants growing in the littoral zone, combined with production from plankton growing in the open water.

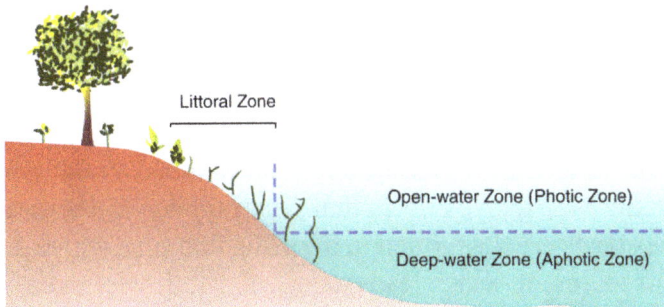

The three primary zones of a lake.

Wetlands can be part of the lentic system, as they form naturally along most lake shores, the width of the wetland and littoral zone being dependent upon the slope of the shoreline and the amount of natural change in water levels, within and among years. Often dead trees accumulate in this zone, either from windfalls on the shore or logs transported to the site during floods. This woody debris provides important habitat for fish and nesting birds, as well as protecting shorelines from erosion.

Two important subclasses of lakes are ponds, which typically are small lakes that intergrade with wetlands, and water reservoirs. Over long periods of time, lakes, or bays within them, may gradually become enriched by nutrients and slowly fill in with organic sediments, a process called succession. When humans use the watershed, the volumes of sediment entering the lake can accelerate this process. The addition of sediments and nutrients to a lake is known as eutrophication.

Ponds

Ponds are small bodies of freshwater with shallow and still water, marsh, and aquatic plants. They can be further divided into four zones: vegetation zone, open water, bottom mud and surface film. The size and depth of ponds often varies greatly with the time of year; many ponds are produced by spring flooding from rivers. Food webs are based both on free-floating algae and upon aquatic plants. There is usually a diverse array of aquatic life, with a few examples including algae, snails, fish, beetles, water bugs, frogs, turtles, otters and muskrats. Top predators may include large fish, herons, or alligators. Since fish are a major predator upon amphibian larvae, ponds that dry up each year, thereby killing resident fish, provide important refugia for amphibian breeding. Ponds that dry up completely each year are often known as vernal pools. Some ponds are produced by animal activity, including alligator holes and beaver ponds, and these add important diversity to landscapes.

Lotic

The major zones in river ecosystems are determined by the river bed's gradient or by the velocity of the current. Faster moving turbulent water typically contains greater concentrations of dissolved oxygen, which supports greater biodiversity than the slow moving water of pools. These distinctions form the basis for the division of rivers into upland and lowland rivers. The food base of streams within riparian forests is mostly derived from the trees, but wider streams and those that lack a canopy derive the majority of their food base from algae. Anadromous fish are also an important source of nutrients. Environmental threats to rivers include loss of water, dams, chemical pollution and introduced species. A dam produces negative effects that continue down the watershed. The most important negative effects are the reduction of spring flooding, which damages wetlands, and the retention of sediment, which leads to loss of deltaic wetlands.

Wetlands

Wetlands are dominated by vascular plants that have adapted to saturated soil. There are four main types of wetlands: swamp, marsh, fen and bog (both fens and bogs are types of mire). Wetlands are the most productive natural ecosystems in the world because of the proximity of water and soil. Hence they support large numbers of plant and animal species. Due to their productivity, wetlands are often converted into dry land with dykes and drains and used for agricultural purposes. The construction of dykes, and dams, has negative consequences for individual wetlands and entire watersheds. Their closeness to lakes and rivers means that they are often developed for human settlement. Once settlements are constructed and protected by dykes, the settlements then become vulnerable to land subsidence and ever increasing risk of flooding. The Louisiana coast around New Orleans is a well-known example; the Danube Delta in Europe is another.

Functions

Aquatic ecosystems perform many important environmental functions. For example, they recycle nutrients, purify water, attenuate floods, recharge ground water and provide habitats for wildlife. Aquatic ecosystems are also used for human recreation, and are very important to the tourism industry, especially in coastal regions.

The health of an aquatic ecosystem is degraded when the ecosystem's ability to absorb a stress has been exceeded. A stress on an aquatic ecosystem can be a result of physical, chemical or biological alterations of the environment. Physical alterations include changes in water temperature, water flow and light availability. Chemical alterations include changes in the loading rates of biostimulatory nutrients, oxygen consuming materials, and toxins. Biological alterations include over-harvesting of commercial species and the introduction of exotic species. Human populations can impose excessive stresses on aquatic ecosystems. There are many examples of excessive stresses with negative consequences. Consider three. The environmental history of the Great Lakes of North America illustrates this problem, particularly how multiple stresses, such as water pollution, over-harvesting and invasive species can combine. The Norfolk Broadlands in England illustrate similar decline with pollution and invasive species. Lake Pontchartrain along the Gulf of Mexico illustrates the negative effects of different stresses including levee construction, logging of swamps, invasive species and salt water intrusion.

Abiotic Characteristics

An ecosystem is composed of biotic communities that are structured by biological interactions and abiotic environmental factors. Some of the important abiotic environmental factors of aquatic ecosystems include substrate type, water depth, nutrient levels, temperature, salinity, and flow. It is often difficult to determine the relative importance of these factors without rather large experiments. There may be complicated feedback loops. For example, sediment may determine the presence of aquatic plants, but aquatic plants may also trap sediment, and add to the sediment through peat.

The amount of dissolved oxygen in a water body is frequently the key substance in determining the extent and kinds of organic life in the water body. Fish need dissolved oxygen to survive, although their tolerance to low oxygen varies among species; in extreme cases of low oxygen some fish even resort to air gulping. Plants often have to produce aerenchyma, while the shape and size of leaves may also be altered. Conversely, oxygen is fatal to many kinds of anaerobic bacteria.

Nutrient levels are important in controlling the abundance of many species of algae. The relative abundance of nitrogen and phosphorus can in effect determine which species of algae come to dominate. Algae are a very important source of food for aquatic life, but at the same time, if they become over-abundant, they can cause declines in fish when they decay. Similar over-abundance of algae in coastal environments such as the Gulf of Mexico produces, upon decay, a hypoxic region of water known as a dead zone.

The salinity of the water body is also a determining factor in the kinds of species found in the water body. Organisms in marine ecosystems tolerate salinity, while many freshwater organisms are intolerant of salt. The degree of salinity in an estuary or delta is an important control upon the type of wetland (fresh, intermediate, or brackish), and the associated animal species. Dams built upstream may reduce spring flooding, and reduce sediment accretion, and may therefore lead to saltwater intrusion in coastal wetlands.

Freshwater used for irrigation purposes often absorbs levels of salt that are harmful to freshwater organisms.

Biotic Characteristics

The biotic characteristics are mainly determined by the organisms that occur. For example, wetland plants may produce dense canopies that cover large areas of sediment—or snails or geese may graze the vegetation leaving large mud flats. Aquatic environments have relatively low oxygen levels, forcing adaptation by the organisms found there. For example, many wetland plants must produce aerenchyma to carry oxygen to roots. Other biotic characteristics are more subtle and difficult to measure, such as the relative importance of competition, mutualism or predation. There are a growing number of cases where predation by coastal herbivores including snails, geese and mammals appears to be a dominant biotic factor.

Autotrophic Organisms

Autotrophic organisms are producers that generate organic compounds from inorganic material. Algae use solar energy to generate biomass from carbon dioxide and are possibly the most important autotrophic organisms in aquatic environments. Of course, the more shallow the water, the greater the biomass contribution from rooted and floating vascular plants. These two sources combine to produce the extraordinary production of estuaries and wetlands, as this autotrophic biomass is converted into fish, birds, amphibians and other aquatic species.

Chemosynthetic bacteria are found in benthic marine ecosystems. These organisms are able to feed on hydrogen sulfide in water that comes from volcanic vents. Great concentrations of animals that feed on these bacteria are found around volcanic vents. For example, there are giant tube worms (*Riftia pachyptila*) 1.5 m in length and clams (*Calyptogena magnifica*) 30 cm long.

Heterotrophic Organisms

Heterotrophic organisms consume autotrophic organisms and use the organic compounds in their bodies as energy sources and as raw materials to create their own biomass. Euryhaline organisms are salt tolerant and can survive in marine ecosystems, while stenohaline or salt intolerant species can only live in freshwater environments.

Freshwater Ecosystem

Freshwater ecosystems are a subset of Earth's aquatic ecosystems. They include lakes and ponds, rivers, streams, springs, and wetlands. They can be contrasted with marine ecosystems, which have a larger salt content. Freshwater habitats can be classified by different factors, including temperature, light penetration, and vegetation.

Freshwater ecosystems can be divided into lentic ecosystems (still water) and lotic ecosystems (flowing water).

Limnology (and its branch freshwater biology) is a study about freshwater ecosystems. It is a part of hydrobiology.

Original efforts to understand and monitor freshwater ecosystems were spurred on by threats to human health (ex. Cholera outbreaks due to sewage contamination). Early monitoring focussed on chemical indicators, then bacteria, and finally algae, fungi and protozoa. A new type of monitoring involves differing groups of organisms (macroinvertebrates, macrophytes and fish) and the stream conditions associated with them.

Current biomonitoring techniques focus mainly on community structure or biochemical oxygen demand. Responses are measured by behavioural changes, altered rates of growth, reproduction or mortality. Macroinvertebrates are most often used in these models because of well known taxonomy, ease of collection, sensitivity to a range of stressors, and their overall value to the ecosystem. Most of these measurements are difficult to extrapolate on a large scale, however.

The use of reference sites is common when assessing what a healthy freshwater ecosystem should "look like". Reference sites are easier to reconstruct in standing water than moving water. Preserved indicators such as diatom valves, macrophyte pollen, insect chitin and fish scales can be used to establish a reference ecosystem representative of a time before large scale human disturbance.

Common chemical stresses on freshwater ecosystem health include acidification, eutrophication and copper and pesticide contamination.

Extinction of Freshwater Fauna

Over 123 freshwater fauna species have gone extinct in North America since 1900. Of North American freshwater species, an estimated 48.5% of mussels, 22.8% of gastropods, 32.7% of crayfishes, 25.9% of amphibians, and 21.3% of fishes are either endangered or threatened. Extinction rates of many species may increase severely into the next century because of invasive species, loss of keystone species, and species which are already functionally extinct. Projected extinction rates for freshwater animals are around five times greater than for land animals, and are comparable to the rates for rainforest communities. Recent extinction trends can be attributed largely to sedimentation, stream fragmentation, chemical and organic pollutants, dams, and invasive species.

Lake Ecosystem

A lake ecosystem includes biotic (living) plants, animals and micro-organisms, as well as abiotic (nonliving) physical and chemical interactions.

Lake ecosystems are a prime example of lentic ecosystems. *Lentic* refers to stationary or relatively still water, from the Latin *lentus*, which means sluggish. Lentic waters range from ponds to lakes to wetlands, and much of this article applies to lentic ecosystems in general. Lentic ecosystems can be compared with lotic ecosystems, which involve flowing terrestrial waters such as rivers and streams. Together, these two fields form the more general study area of freshwater or aquatic ecology.

Lentic systems are diverse, ranging from a small, temporary rainwater pool a few inches deep to Lake Baikal, which has a maximum depth of 1740 m. The general distinction between pools/ponds and lakes is vague, but Brown states that ponds and pools have their entire bottom surfaces exposed to light, while lakes do not. In addition, some lakes become seasonally stratified (discussed in more detail below.) Ponds and pools have two regions: the pelagic open water zone, and the benthic zone, which comprises the bottom and shore regions. Since lakes have deep bottom regions not exposed to light, these systems have an additional zone, the profundal. These three areas can have very different abiotic conditions and, hence, host species that are specifically adapted to live there.

Important Abiotic Factors

Light

Light provides the solar energy required to drive the process of photosynthesis, the major energy source of lentic systems. The amount of light received depends upon a combination of several factors. Small ponds may experience shading by surrounding trees, while cloud cover may affect light availability in all systems, regardless of size. Seasonal and diurnal considerations also play a role in light availability because the shallower the angle at which light strikes water, the more light is lost by reflection. This is known as Beer's law. Once light has penetrated the surface, it may also be scattered by particles suspended in the water column. This scattering decreases the total amount of light as depth increases. Lakes are divided into photic and aphotic regions, the prior receiving sunlight and latter being below the depths of light penetration, making it void of photosynthetic capacity. In relation to lake zonation, the pelagic and benthic zones are considered to lie within the photic region, while the profundal zone is in the aphotic region.

Temperature

Temperature is an important abiotic factor in lentic ecosystems because most of the biota are poikilothermic, where internal body temperatures are defined by the surrounding system. Water can be heated or cooled through radiation at the surface and conduction to or from the air and surrounding substrate. Shallow ponds often have a continuous temperature gradient from warmer waters at the surface to cooler waters at the bottom. In addition, temperature fluctuations can be very great in these systems, both diurnally and seasonally.

Temperature regimes are very different in large lakes (Fig. 2). In temperate regions, for example, as air temperatures increase, the icy layer formed on the surface of the lake breaks up, leaving the water at approximately 4 °C. This is the temperature at which water has the highest density. As the season progresses, the warmer air temperatures heat the surface waters, making them less dense. The deeper waters remain cool and dense due to reduced light penetration. As the summer begins, two distinct layers become established, with such a large temperature difference between them that they remain stratified. The lowest zone in the lake is the coldest and is called the hypolimnion. The upper warm zone is called the epilimnion. Between these zones is a band of rapid temperature change called the thermocline. During the colder fall season, heat is lost at the surface and the epilimnion cools. When the temperatures of the two zones are close enough, the waters begin to mix again to create a uniform temperature, an event termed lake turnover. In the winter, inverse stratification occurs as water near the surface cools freezes, while warmer, but denser water remains near the bottom. A thermocline is established, and the cycle repeats.

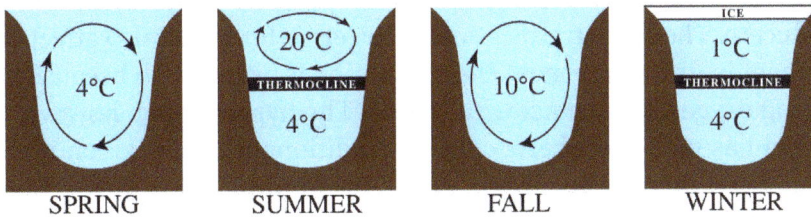

Fig. 2 Seasonal stratification in temperate lakes

Wind

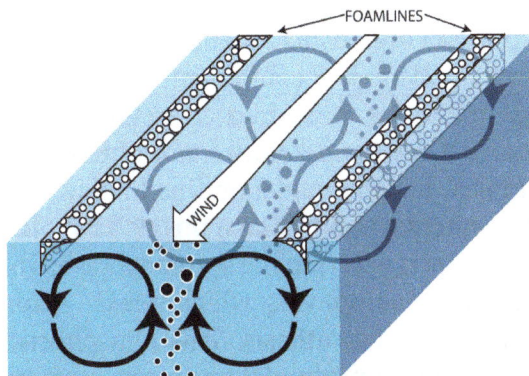

Fig. 3 Illustration of Langmuir rotations; open circles=positively buoyant particles, closed circles=negatively buoyant particles

In exposed systems, wind can create turbulent, spiral-formed surface currents called Langmuir circulations (Fig. 3). Exactly how these currents become established is still not well understood, but it is evident that it involves some interaction between horizontal surface currents and surface gravity waves. The visible result of these rotations, which can be seen in any lake, are the surface foamlines that run parallel to the wind direction. Positively buoyant particles and small organisms concentrate in the foamline at the surface and negatively buoyant objects are found in the upwelling current between the two rotations. Objects with neutral buoyancy tend to be evenly distributed in the water column. This turbulence circulates nutrients in the water column, making it crucial for many pelagic species, however its effect on benthic and profundal organisms is minimal to non-existent, respectively. The degree of nutrient circulation is system specific, as it depends upon such factors as wind strength and duration, as well as lake or pool depth and productivity.

Chemistry

Oxygen is essential for organismal respiration. The amount of oxygen present in standing waters depends upon: 1) the area of transparent water exposed to the air, 2) the circulation of water within the system and 3) the amount of oxygen generated and used by organisms present. In shallow, plant-rich pools there may be great fluctuations of oxygen, with extremely high concentrations occurring during the day due to photosynthesis and very low values at night when respiration is the dominant process of primary producers. Thermal stratification in larger systems can also affect the amount of oxygen present in different zones. The epilimnion is oxygen rich because it circulates quickly, gaining oxygen via contact with the air. The hypolimnion, however, circulates very slowly and has no atmospheric contact. Additionally, fewer green plants exist in the hypolimnion, so there is less oxygen released from photosynthesis. In spring and fall when the epilimnion and hypolimnion mix, oxygen becomes more evenly distributed in the system. Low oxygen levels are characteristic of the profundal zone due to the accumulation of decaying vegetation and animal matter that "rains" down from the pelagic and benthic zones and the inability to support primary producers.

Phosphorus is important for all organisms because it is a component of DNA and RNA and is involved in cell metabolism as a component of ATP and ADP. Also, phosphorus is not found in large quantities in freshwater systems, limiting photosynthesis in primary producers, making it the main determinant of lentic system production. The phosphorus cycle is complex, but the model outlined below describes the basic pathways. Phosphorus mainly enters a pond or lake through runoff from the watershed or by atmospheric deposition. Upon entering the system, a reactive form of phosphorus is usually taken up by algae and macrophytes, which release a non-reactive phosphorus compound as a byproduct of photosynthesis. This phosphorus can drift downwards and become part of the benthic or profundal sediment, or it can be remineralized to the reactive form by microbes in the water column. Similarly, non-reactive phosphorus in the sediment can be remineralized into the reactive form. Sediments are generally richer in phosphorus than lake water, however, indicating that this nutrient may have a long residency time there before it is remineralized and re-introduced to the system.

Lentic System Biota

Bacteria

Bacteria are present in all regions of lentic waters. Free-living forms are associated with decomposing organic material, biofilm on the surfaces of rocks and plants, suspended in the water column, and in the sediments of the benthic and profundal zones. Other forms are also associated with the guts of lentic animals as parasites or in commensal relationships. Bacteria play an important role in system metabolism through nutrient recycling, which is discussed in the Trophic Relationships section.

Primary Producers

Nelumbo nucifera, an aquatic plant.

Algae, including both phytoplankton and periphyton are the principle photosynthe-sizers in ponds and lakes. Phytoplankton are found drifting in the water column of the pelagic zone. Many species have a higher density than water which should make them sink and end up in the benthos. To combat this, phytoplankton have developed density changing mechanisms, by forming vacuoles and gas vesicles or by changing their shapes to induce drag, slowing their descent. A very sophisticated adaptation utilized by a small number of species is a tail-like flagellum that can adjust vertical position and allow movement in any direction. Phytoplankton can also maintain their presence in the water column by being circulated in Langmuir rotations. Periphytic algae, on the other hand, are attached to a substrate. In lakes and ponds, they can cover all benthic surfaces. Both types of plankton are important as food sources and as oxygen providers.

Aquatic plants live in both the benthic and pelagic zones and can be grouped according to their manner of growth: 1) emergent = rooted in the substrate but with leaves and flowers extending into the air, 2) floating-leaved = rooted in the substrate but with floating leaves, 3) submersed = growing beneath the surface and 4) free-floating macrophytes = not rooted in the substrate and floating on the surface. These various forms of macrophytes generally occur in different areas of the benthic zone, with emergent vegetation nearest the shoreline, then floating-leaved macrophytes, followed by submersed vegetation. Free-floating macrophytes can occur anywhere on the system's surface.

Aquatic plants are more buoyant than their terrestrial counterparts because freshwater has a higher density than air. This makes structural rigidity unimportant in lakes and ponds (except in the aerial stems and leaves). Thus, the leaves and stems of most aquatic plants use less energy to construct and maintain woody tissue, investing that energy into fast growth instead. In order to contend with stresses induced by wind and waves, plants must be both flexible and tough. Light, water depth and substrate types are the most important factors controlling the distribution of submerged aquatic plants. Macrophytes are sources of food, oxygen, and habitat structure in the benthic zone, but cannot penetrate the depths of the euphotic zone and hence are not found there.

Invertebrates

Water striders are predatory insects which rely on surface tension to walk on top of water. They live on the surface of ponds, marshes, and other quiet waters. They can move very quickly, up to 1.5 m/s.

Zooplankton are tiny animals suspended in the water column. Like phytoplankton, these species have developed mechanisms that keep them from sinking to deeper waters, including drag-inducing body forms and the active flicking of appendages such as antennae or spines. Remaining in the water column may have its advantages in terms of feeding, but this zone's lack of refugia leaves zooplankton vulnerable to predation. In response, some species, especially Daphnia sp., make daily vertical migrations in the water column by passively sinking to the darker lower depths during the day and actively moving towards the surface during the night. Also, because conditions in a lentic system can be quite variable across seasons, zooplankton have the ability to switch from laying regular eggs to resting eggs when there is a lack of food, temperatures fall below 2 °C, or if predator abundance is high. These resting eggs have a diapause, or dormancy period that should allow the zooplankton to encounter conditions that are more favorable to survival when they finally hatch. The invertebrates that inhabit the benthic zone are numerically dominated by small species and are species rich compared to the zooplankton of the open water. They include Crustaceans (e.g. crabs, crayfish, and shrimp), molluscs (e.g. clams and snails), and numerous types of insects. These organisms are mostly found in the areas of macrophyte growth, where the richest resources, highly oxygenated water, and warmest portion of the ecosystem are found. The structurally diverse macrophyte beds are important sites for the accumulation of organic matter, and provide an ideal area for colonization. The sediments and plants also offer a great deal of protection from predatory fishes.

Very few invertebrates are able to inhabit the cold, dark, and oxygen poor profundal zone. Those that can are often red in color due to the presence of large amounts of hemoglobin, which greatly increases the amount of oxygen carried to cells. Because the concentration of oxygen within this zone is low, most species construct tunnels or borrows in which they can hide and make the minimum movements necessary to circulate water through, drawing oxygen to them without expending much energy.

Fish and Other Vertebrates

Fish have a range of physiological tolerances that are dependent upon which species they belong to. They have different lethal temperatures, dissolved oxygen requirements, and spawning needs that are based on their activity levels and behaviors. Because fish are highly mobile, they are able to deal with unsuitable abiotic factors in one zone by simply moving to another. A detrital feeder in the profundal zone, for example, that finds the oxygen concentration has dropped too low may feed closer to the benthic zone. A fish might also alter its residence during different parts of its life history: hatching in a sediment nest, then moving to the weedy benthic zone to develop in a protected environment with food resources, and finally into the pelagic zone as an adult.

Other vertebrate taxa inhabit lentic systems as well. These include amphibians (e.g. salamanders and frogs), reptiles (e.g. snakes, turtles, and alligators), and a large number of waterfowl species. Most of these vertebrates spend part of their time in terrestrial habitats and thus are not directly affected by abiotic factors in the lake or pond. Many fish species are important as consumers and as prey species to the larger vertebrates mentioned above.

Trophic Relationships

Primary Producers

Lentic systems gain most of their energy from photosynthesis performed by aquatic plants and algae. This autochthonous process involves the combination of carbon dioxide, water, and solar energy to produce carbohydrates and dissolved oxygen. Within a lake or pond, the potential rate of photosynthesis generally decreases with depth due to light attenuation. Photosynthesis, however, is often low at the top few millimeters of the surface, likely due to inhibition by ultraviolet light. The exact depth and photosynthetic rate measurements of this curve are system specific and depend upon: 1) the total biomass of photosynthesizing cells, 2) the amount of light attenuating materials and 3) the abundance and frequency range of light absorbing pigments (i.e. chlorophylls) inside of photosynthesizing cells. The energy created by these primary producers is important for the community because it is transferred to higher trophic levels via consumption.

Bacteria

The vast majority of bacteria in lakes and ponds obtain their energy by decomposing vegetation and animal matter. In the pelagic zone, dead fish and the occasional allochthonous input of litterfall are examples of coarse particulate organic matter (CPOM>1 mm). Bacteria degrade these into fine particulate organic matter (FPOM<1 mm) and then further into usable nutrients. Small organisms such as plankton are also characterized as FPOM. Very low concentrations of nutrients are released during decomposition because the bacteria are utilizing them to build their own biomass. Bacteria, however, are consumed by protozoa, which are in turn consumed by zooplankton, and then further up the trophic levels. Nutrients, including those that contain carbon and phosphorus, are reintroduced into the water column at any number of points along this food chain via excretion or organism death, making them available again for bacteria. This regeneration cycle is known as the microbial loop and is a key component of lentic food webs.

The decomposition of organic materials can continue in the benthic and profundal zones if the matter falls through the water column before being completely digested by the pelagic bacteria. Bacteria are found in the greatest abundance here in sediments, where they are typically 2-1000 times more prevalent than in the water column.

Benthic Invertebrates

Benthic invertebrates, due to their high level of species richness, have many methods of prey capture. Filter feeders create currents via siphons or beating cilia, to pull water and its nutritional contents, towards themselves for straining. Grazers use scraping, rasping, and shredding adaptations to feed on periphytic algae and macrophytes. Members of the collector guild browse the sediments, picking out specific particles with raptorial appendages. Deposit feeding invertebrates indiscriminately consume sediment, digesting any organic material it contains. Finally, some invertebrates belong to the predator guild, capturing and consuming living animals. The profundal zone is home to a unique group of filter feeders that use small body movements to draw a current through burrows that they have created in the sediment. This mode of feeding requires the least amount of motion, allowing these species to conserve energy. A small number of invertebrate taxa are predators in the profundal zone. These species are likely from other regions and only come to these depths to feed. The vast majority of invertebrates in this zone are deposit feeders, getting their energy from the surrounding sediments.

Fish

Fish size, mobility, and sensory capabilities allow them to exploit a broad prey base, covering multiple zonation regions. Like invertebrates, fish feeding habits can be categorized into guilds. In the pelagic zone, herbivores graze on periphyton and macrophytes or pick phytoplankton out of the water column. Carnivores include fishes that feed on zooplankton in the water column (zooplanktivores), insects at the water's surface, on benthic structures, or in the sediment (insectivores), and those that feed on other fish (piscivores). Fish that consume detritus and gain energy by processing its organic material are called detritivores. Omnivores ingest a wide variety of prey, encompassing floral, faunal, and detrital material. Finally, members of the parasitic guild acquire nutrition from a host species, usually another fish or large vertebrate. Fish taxa are flexible in their feeding roles, varying their diets with environmental conditions and prey availability. Many species also undergo a diet shift as they develop. Therefore, it is likely that any single fish occupies multiple feeding guilds within its lifetime.

Lentic Food Webs

As noted in the previous sections, the lentic biota are linked in complex web of trophic relationships. These organisms can be considered to loosely be associated with specific trophic groups (e.g. primary producers, herbivores, primary carnivores, secondary carnivores, etc.). Scientists have developed several theories in order to understand the mechanisms that control the abundance and diversity within these groups. Very generally, top-down processes dictate that the abundance of prey taxa is dependent upon the actions of consumers from higher trophic levels. Typically, these processes operate only between two trophic levels, with no effect on the others. In some cases, however, aquatic systems experience a trophic cascade; for example, this might occur if primary producers experience less grazing by herbivores because these herbivores are suppressed by carnivores. Bottom-up processes are functioning when the abundance or diversity of members of higher trophic levels is dependent upon the availability or quality of resources from lower levels. Finally, a combined regulating theory, bottom-up:top-down, combines the predicted influences of consumers and resource availability. It predicts that trophic levels close to the lowest trophic levels will be most influenced by bottom-up forces, while top-down effects should be strongest at top levels.

Community Patterns and Diversity

Local Species Richness

The biodiversity of a lentic system increases with the surface area of the lake or pond. This is attributable to the higher likelihood of partly terrestrial species of finding a larger system. Also, because larger systems typically have larger populations, the chance of extinction is decreased. Additional factors, including temperature regime, pH, nutrient availability, habitat complexity, speciation rates, competition, and predation, have been linked to the number of species present within systems.

Succession patterns in plankton communities – the PEG model

Phytoplankton and zooplankton communities in lake systems undergo seasonal succession in relation to nutrient availability, predation, and competition. Sommer *et al.* described these patterns as part of the Plankton Ecology Group (PEG) model, with 24 statements constructed from the analysis of numerous systems. The following includes a subset of these statements, as explained by Brönmark and Hansson illustrating succession through a single seasonal cycle:

Winter 1. Increased nutrient and light availability result in rapid phytoplankton growth towards the end of winter. The dominant species, such as diatoms, are small and have quick growth capabilities. 2. These plankton are consumed by zooplankton, which become the dominant plankton taxa.

Spring 3. A clear water phase occurs, as phytoplankton populations become depleted due to increased predation by growing numbers of zooplankton.

Summer 4. Zooplankton abundance declines as a result of decreased phytoplankton prey and increased predation by juvenile fishes. 5. With increased nutrient availability and decreased predation from zooplankton, a diverse phytoplankton community develops. 6. As the summer continues, nutrients become depleted in a predictable order: phosphorus, silica, and then nitrogen. The abundance of various phytoplankton species varies in relation to their biological need for these nutrients. 7. Small-sized zooplankton become the dominant type of zooplankton because they are less vulnerable to fish predation.

Fall 8. Predation by fishes is reduced due to lower temperatures and zooplankton of all sizes increase in number.

Winter 9. Cold temperatures and decreased light availability result in lower rates of primary production and decreased phytoplankton populations. 10. Reproduction in zooplankton decreases due to lower temperatures and less prey.

The PEG model presents an idealized version of this succession pattern, while natural systems are known for their variation.

Latitudinal Patterns

There is a well-documented global pattern that correlates decreasing plant and animal diversity with increasing latitude, that is to say, there are fewer species as one moves towards the poles. The cause of this pattern is one of the greatest puzzles for ecologists today. Theories for its explanation include energy availability, climatic variability, disturbance, competition, etc. Despite this global diversity gradient, this pattern can be weak for freshwater systems compared to global marine and terrestrial systems. This may be related to size, as Hillebrand and Azovsky found that smaller organisms (protozoa and plankton) did not follow the expected trend strongly, while larger species (vertebrates) did. They attributed this to better dispersal ability by smaller organisms, which may result in high distributions globally.

Natural Lake Lifecycles

Lake Creation

Lakes can be formed in a variety of ways, but the most common are discussed briefly below. The oldest and largest systems are the result of tectonic activities. The rift lakes in Africa, for example are the result of seismic activity along the site of separation of two tectonic plates. Ice-formed lakes are created when glaciers recede, leaving behind abnormalities in the landscape shape that are then filled with water. Finally, oxbow lakes are fluvial in origin, resulting when a meandering river bend is pinched off from the main channel.

Natural Extinction

All lakes and ponds receive sediment inputs. Since these systems are not really expanding, it is logical to assume that they will become increasingly shallower in depth, eventually becoming wetlands or terrestrial vegetation. The length of this process should depend upon a combination of depth and sedimentation rate. Moss gives the example of Lake Tanganyika, which reaches a depth of 1500 m and has a sedimentation rate of 0.5 mm/yr. Assuming that sedimentation is not influenced by anthropogenic factors, this system should go extinct in approximately 3 million years. Shallow lentic systems might also fill in as swamps encroach inward from the edges. These processes operate on a much shorter timescale, taking hundreds to thousands of years to complete the extinction process.

Human Impacts

Acidification

Sulfur dioxide and nitrogen oxides are naturally released from volcanoes, organic compounds in the soil, wetlands, and marine systems, but the majority of these compounds come from the combustion of coal, oil, gasoline, and the smelting of ores containing sulfur. These substances dissolve in atmospheric moisture and enter lentic systems as acid rain. Lakes and ponds that contain bedrock that is rich in carbonates have a natural buffer, resulting in no alteration of pH. Systems without this bedrock, however, are very sensitive to acid inputs because they have a low neutralizing capacity, resulting in pH declines even with only small inputs of acid. At a pH of 5–6 algal species diversity and biomass decrease considerably, leading to an increase in water transparency – a characteristic feature of acidified lakes. As the pH continues lower, all fauna becomes less diverse. The most significant feature is the disruption of fish reproduction. Thus, the population is eventually composed of few, old individuals that eventually die and leave the systems without fishes. Acid rain has been especially harmful to lakes in Scandinavia, western Scotland, west Wales and the north eastern United States.

Eutrophication

Eutrophic systems contain a high concentration of phosphorus (~30 µg/L), nitrogen (~1500 µg/L), or both. Phosphorus enters lentic waters from sewage treatment effluents, discharge from raw sewage, or from runoff of farmland. Nitrogen mostly comes from agricultural fertilizers from runoff or leaching and subsequent groundwater flow. This increase in nutrients required for primary producers results in a massive increase of phytoplankton growth, termed a plankton bloom. This bloom decreases water transparency, leading to the loss of submerged plants. The resultant reduction in habitat structure has negative impacts on the species' that utilize it for spawning, maturation and general survival. Additionally, the large number of short-lived phytoplankton result in a massive amount of dead biomass settling into the sediment. Bacteria need large amounts of oxygen to decompose this material, reducing the oxygen concentration of the water. This is especially pronounced in stratified lakes when the thermocline prevents oxygen rich water from the surface to mix with lower levels. Low or anoxic conditions preclude the existence of many taxa that are not physiologically tolerant of these conditions.

Invasive Species

Invasive species have been introduced to lentic systems through both purposeful events (e.g. stocking game and food species) as well as unintentional events (e.g. in ballast water). These organisms can affect natives via competition for prey or habitat, predation, habitat alteration, hybridization, or the introduction of harmful diseases and parasites. With regard to native species, invaders may cause changes in size and age structure, distribution, density, population growth, and may even drive populations to extinction. Examples of prominent invaders of lentic systems include the zebra mussel and sea lamprey in the Great Lakes.

River Ecosystem

The ecosystem of a river is the river viewed as a system operating in its natural environment, and includes biotic (living) interactions amongst plants, animals and micro-organisms, as well as abiotic (nonliving) physical and chemical interactions.

River ecosystems are prime examples of lotic ecosystems. *Lotic* refers to flowing water, from the Latin *lotus*, washed. Lotic waters range from springs only a few centimeters wide to major rivers kilometers in width. Much of this article applies to lotic ecosystems in general, including related lotic systems such as streams and springs. Lotic ecosystems can be contrasted with lentic ecosystems, which involve relatively still terrestrial waters such as lakes and ponds. Together, these two fields form the more general study area of freshwater or aquatic ecology.

This stream in the redwoods together with its environment can be thought of as forming a river or lotic ecosystem

The following unifying characteristics make the ecology of running waters unique from that of other aquatic habitats.

- Flow is unidirectional.

- There is a state of continuous physical change.

- There is a high degree of spatial and temporal heterogeneity at all scales (microhabitats).

- Variability between lotic systems is quite high.

- The biota is specialized to live with flow conditions.

Abiotic Factors

The non living components of an ecosystem are called abiotic components

Rapids in Mount Robson Provincial Park

River

Water flow is the key factor in lotic systems influencing their ecology. The strength of water flow can vary between systems, ranging from torrential rapids to slow backwaters that almost seem like lentic systems. The speed of the water flow can also vary within a system and is subject to chaotic turbulence. This turbulence results in divergences of flow from the mean downslope flow vector as typified by eddy currents. The mean flow rate vector is based on variability of friction with the bottom or sides of the channel, sinuosity, obstructions, and the incline gradient. In addition, the amount of water input into the system from direct precipitation, snowmelt, and/or groundwater can affect flow rate. Flowing waters can alter the shape of the streambed through erosion and deposition, creating a variety of habitats, including riffles, glides, and pools.

Light

A pensive Cooplacurripa River, NSW

Light is important to lotic systems, because it provides the energy necessary to drive primary production via photosynthesis, and can also provide refuge for prey species in shadows it casts. The amount of light that a system receives can be related to a combination of internal and external stream variables. The area surrounding a small stream, for example, might be shaded by surrounding forests or by valley walls. Larger river systems tend to be wide so the influence of external variables is minimized, and the sun reaches the surface. These rivers also tend to be more turbulent, however, and particles in the water increasingly attenuate light as depth increases. Seasonal and diurnal factors might also play a role in light availability because the angle of incidence, the angle at which light strikes water can lead to light lost from reflection. Known as Beer's Law, the shallower the angle, the more light is reflected and the amount of solar radiation received declines logarithmically with depth. Additional influences on light availability include cloud cover, altitude, and geographic position (Brown 1987).

Temperature

Castle Geyser, Yellowstone National Park

Most lotic species are poikilotherms whose internal temperature varies with their environment, thus temperature is a key abiotic factor for them. Water can be heated or cooled through radiation at the surface and conduction to or from the air and surrounding substrate. Shallow streams are typically well mixed and maintain a relatively uniform temperature within an area. In deeper, slower moving water systems, however, a strong difference between the bottom and surface temperatures may develop. Spring fed systems have little variation as springs are typically from groundwater sources, which are often very close to ambient temperature. Many systems show strong diurnal fluctuations and seasonal variations are most extreme in arctic, desert and temperate systems. The amount of shading, climate and elevation can also influence the temperature of lotic systems.

Chemistry

A forest stream in the winter near Erzhausen, Germany

Water chemistry between systems varies tremendously. The chemistry is foremost determined by inputs from the geology of its watershed, or catchment area, but can also be influenced by precipitation and the addition of pollutants from human sources. Large differences in chemistry do not usually exist within small lotic systems due to a high rate of mixing. In larger river systems, however, the concentrations of most nutrients, dissolved salts, and pH decrease as distance increases from the river's source.

Oxygen is likely the most important chemical constituent of lotic systems, as all aerobic organisms require it for survival. It enters the water mostly via diffusion at the water-air interface. Oxygen's solubility in water decreases as water pH and temperature increases. Fast, turbulent streams expose more of the water's surface area to the air and tend to have low temperatures and thus more oxygen than slow, backwaters. Oxygen is a by-product of photosynthesis, so systems with a high abundance of aquatic algae and plants may also have high concentrations of oxygen during the day. These levels can decrease significantly during the night when primary producers switch to respiration. Oxygen can be limiting if circulation between the surface and deeper layers is poor, if the activity of lotic animals is very high, or if there is a large amount of organic decay occurring.

Substrate

Cascade in the Pyrénées.

The inorganic substrate of lotic systems is composed of the geologic material present in the catchment that is eroded, transported, sorted, and deposited by the current. Inorganic substrates are classified by size on the Wentworth scale, which ranges from boulders, to pebbles, to gravel, to sand, and to silt. Typically, particle size decreases downstream with larger boulders and stones in more mountainous areas and sandy bottoms in lowland rivers. This is because the higher gradients of mountain streams facilitate a faster flow, moving smaller substrate materials further downstream for deposition. Substrate can also be organic and may include fine particles, autumn shed leaves, submerged wood, moss, and more evolved plants. Substrate deposition is not necessarily a permanent event, as it can be subject to large modifications during flooding events.

Biotic Factors

The living components of an ecosystem are called the biotic components.

Bacteria

Bacteria are present in large numbers in lotic waters. Free-living forms are associated with decomposing organic material, biofilm on the surfaces of rocks and vegetation, in between particles that compose the substrate, and suspended in the water column. Other forms are also associated with the guts of lotic organisms as parasites or in commensal relationships. Bacteria play a large role in energy recycling, which will be discussed in the Trophic Relationships section.

Primary Producers

Periphyton

Common water hyacinth in flower

Algae, consisting of phytoplankton and periphyton, are the most significant sources of primary production in most streams and rivers. Phytoplankton float freely in the water column and thus are unable to maintain populations in fast flowing streams. They can, however, develop sizable populations in slow moving rivers and backwaters. Periphyton are typically filamentous and tufted algae that can attach themselves to objects to avoid being washed away by fast currents. In places where flow rates are negligible or absent, periphyton may form a gelatinous, unanchored floating mat.

Plants exhibit limited adaptations to fast flow and are most successful in reduced currents. More primitive plants, such as mosses and liverworts attach themselves to solid objects. This typically occurs in colder headwaters where the mostly rocky substrate offers attachment sites. Some plants are free floating at the water's surface in dense mats like duckweed or water hyacinth. Others are rooted and may be classified as submerged or emergent. Rooted plants usually occur in areas of slackened current where fine-grained soils are found (Brown 1987). These rooted plants are flexible, with elongated leaves that offer minimal resistance to current.

Living in flowing water can be beneficial to plants and algae because the current is usually well aerated and it provides a continuous supply of nutrients. These organisms are limited by flow, light, water chemistry, substrate, and grazing pressure. Algae and plants are important to lotic systems as sources of energy, for forming microhabitats that shelter other fauna from predators and the current, and as a food resource (Brown 1987).

Insects and Other Invertebrates

Up to 90% of invertebrates in some lotic systems are insects. These species exhibit tremendous diversity and can be found occupying almost every available habitat, including the surfaces of stones, deep below the substratum, adrift in the current, and in the surface film. Insects have developed several strategies for living in the diverse flows of lotic systems. Some avoid high current areas, inhabiting the substratum or the sheltered side of rocks. Additional invertebrate taxa common to flowing waters include mollusks such as snails, limpets, clams, mussels, as well as crustaceans like crayfish and crabs. Like most of the primary consumers, lotic invertebrates often rely heavily on the current to bring them food and oxygen (Brown 1987). Invertebrates, especially insects, are important as both consumers and prey items in lotic systems.

Fish and Other Vertebrates

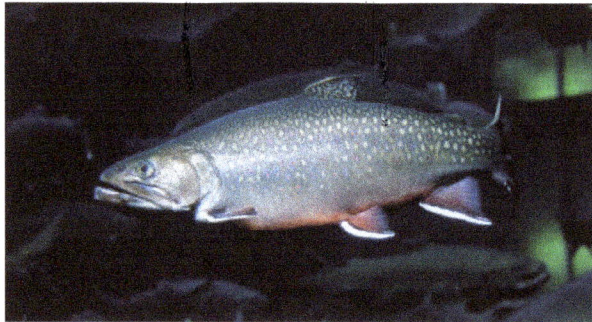

The brook trout is native to small streams, creeks, lakes, and spring ponds.

Fish are probably the best-known inhabitants of lotic systems. The ability of a fish species to live in flowing waters depends upon the speed at which it can swim and the duration that its speed can be maintained. This ability can vary greatly between species and is tied to the habitat in which it can survive. Continuous swimming expends a tremendous amount of energy and, therefore, fishes spend only short periods in full current. Instead, individuals remain close to the bottom or the banks, behind obstacles, and sheltered from the current, swimming in the current only to feed or change locations. Some species have adapted to living only on the system bottom, never venturing into the open water flow. These fishes are dorso-ventrally flattened to reduce flow resistance and often have eyes on top of their heads to observe what is happening above them. Some also have sensory barrels positioned under the head to assist in the testing of substratum (Brown 1987).

New Zealand longfin eels can weigh over 50 kilograms.

Lotic systems typically connect to each other, forming a path to the ocean (spring → stream → river → ocean), and many fishes have life cycles that require stages in both fresh and salt water. Salmon, for example, are anadromous species that are born in freshwater but spend most of their adult life in the ocean, returning to fresh water only to spawn. Eels are catadromous species that do the opposite, living in freshwater as adults but migrating to the ocean to spawn.

Other vertebrate taxa that inhabit lotic systems include amphibians, such as salamanders, reptiles (e.g. snakes, turtles, crocodiles and alligators) various bird species, and mammals (e.g., otters, beavers, hippos, and river dolphins). With the exception of a few species, these vertebrates are not tied to water as fishes are, and spend part of their time in terrestrial habitats. Many fish species are important as consumers and as prey species to the larger vertebrates mentioned above.

Trophic Relationships

Energy Inputs

Pondweed is an autochthonous energy source

Leaf litter is an allochthonous energy source

Energy sources can be autochthonous or allochthonous.

- Autochthonous energy sources are those derived from within the lotic system. During photosynthesis, for example, primary producers form organic carbon compounds out of carbon dioxide and inorganic matter. The energy they produce is important for the community because it may be transferred to higher trophic levels via consumption. Additionally, high rates of primary production can introduce dissolved organic matter (DOM) to the waters. Another form of autochthonous energy comes from the decomposition of dead organisms and feces that originate within the lotic system. In this case, bacteria decompose the detritus or coarse particulate organic material (CPOM; >1 mm pieces) into fine particulate organic matter (FPOM; <1 mm pieces) and then further into inorganic compounds that are required for photosynthesis. This process is discussed in more detail below.

- Allochthonous energy sources are those derived from outside the lotic system, that is, from the terrestrial environment. Leaves, twigs, fruits, etc. are typical forms of terrestrial CPOM that have entered the water by direct litterfall or lateral leaf blow. In addition, terrestrial animal-derived materials, such as feces or carcasses that have been added to the system are examples of allochthonous CPOM. The CPOM undergoes a specific process of degradation. Allan gives the example of a leaf fallen into a stream. First, the soluble chemicals are dissolved and leached from the leaf upon its saturation with water. This adds to the DOM load in the system. Next, microbes such as bacteria and fungi colonize the leaf, softening it as the mycelium of the fungus grows into it. The composition of the microbial community is influenced by the species of tree from which the leaves are shed (Rubbo and Kiesecker 2004). This combination of bacteria, fungi, and leaf are a food source for shredding invertebrates, which leave only FPOM after consumption. These fine particles may be colonized by microbes again or serve as a food source for animals that consume FPOM. Organic matter can also enter the lotic system already in the FPOM stage by wind, surface runoff, bank erosion, or groundwater. Similarly, DOM can be introduced through canopy drip from rain or from surface flows.

Invertebrates

Invertebrates can be organized into many feeding guilds in lotic systems. Some species are shredders, which use large and powerful mouth parts to feed on non-woody CPOM and their associated microorganisms. Others are suspension feeders, which use their setae, filtering aparati, nets, or even secretions to collect FPOM and microbes from the water. These species may be passive collectors, utilizing the natural flow of the system, or they may generate their own current to draw water, and also, FPOM in Allan. Members of the gatherer-collector guild actively search for FPOM under rocks and in other places where the stream flow has slackened enough to allow deposition. Grazing invertebrates utilize scraping, rasping, and browsing adaptations to feed on periphyton and detritus. Finally, several families are predatory, capturing and consuming animal prey. Both the number of species and the abundance of individuals within each guild is largely dependent upon food availability. Thus, these values may vary across both seasons and systems.

Fish

Fish can also be placed into feeding guilds. Planktivores pick plankton out of the water column. Herbivore-detritivores are bottom-feeding species that ingest both periphyton and detritus indiscriminately. Surface and water column feeders capture surface prey (mainly terrestrial and emerging insects) and drift (benthic invertebrates floating downstream). Benthic invertebrate feeders prey primarily on immature insects, but will also consume other benthic invertebrates. Top predators consume fishes and/or large invertebrates. Omnivores ingest a wide range of prey. These can be floral, faunal, and/or detrital in nature. Finally, parasites live off of host species, typically other fishes. Fish are flexible in their feeding roles, capturing different prey with regard to seasonal availability and their own developmental stage. Thus, they may occupy multiple feeding guilds in their lifetime. The number of species in each guild can vary greatly between systems, with temperate warm water streams having the most benthic invertebrate feeders, and tropical systems having large numbers of detritus feeders due to high rates of allochthonous input.

Local Species Richness

Large rivers have comparatively more species than small streams. Many relate this pattern to the greater area and volume of larger systems, as well as an increase in habitat diversity. Some systems, however, show a poor fit between system size and species richness. In these cases, a combination of factors such as historical rates of speciation and extinction, type of substrate, microhabitat availability, water chemistry, temperature, and disturbance such as flooding seem to be important.

Resource Partitioning

Although many alternate theories have been postulated for the ability of guild-mates to coexist, resource partitioning has been well documented in lotic systems as a means of reducing competition. The three main types of resource partitioning include habitat, dietary, and temporal segregation.

Habitat segregation was found to be the most common type of resource partitioning in natural systems (Schoener, 1974). In lotic systems, microhabitats provide a level of physical complexity that can support a diverse array of organisms (Vincin and Hawknis, 1998). The separation of species by substrate preferences has been well documented for invertebrates. Ward (1992) was able to divide substrate dwellers into six broad assemblages, including those that live in: coarse substrate, gravel, sand, mud, woody debris, and those associated with plants, showing one layer of segregation. On a smaller scale, further habitat partitioning can occur on or around a single substrate, such as a piece of gravel. Some invertebrates prefer the high flow areas on the exposed top of the gravel, while others reside in the crevices between one piece of gravel and the next, while still others live on the bottom of this gravel piece.

Dietary segregation is the second-most common type of resource partitioning. High degrees of morphological specializations or behavioral differences allow organisms to use specific resources. The size of nets built by some species of invertebrate suspension feeders, for example, can filter varying particle size of FPOM from the water (Edington et al. 1984). Similarly, members in the grazing guild can specialize in the harvesting of algae or detritus depending upon the morphology of their scraping apparatus. In addition, certain species seem to show a preference for specific algal species.

Temporal segregation is a less common form of resource partitioning, but it is nonetheless an observed phenomenon. Typically, it accounts for coexistence by relating it to differences in life history patterns and the timing of maximum growth among guild mates. Tropical fishes in Borneo, for example, have shifted to shorter life spans in response to the ecological niche reduction felt with increasing levels of species richness in their ecosystem (Watson and Balon 1984).

Persistence and Succession

Over long time scales, there is a tendency for species composition in pristine systems to remain in a stable state. This has been found for both invertebrate and fish species. On shorter time scales, however, flow variability and unusual precipitation patterns decrease habitat stability and can all lead to declines in persistence levels. The ability to maintain this persistence over long time scales is related to the ability of lotic systems to return to the original community configuration relatively quickly after a disturbance (Townsend et al. 1987). This is one example of temporal succession, a site-specific change in a community involving changes in species composition over time. Another form of temporal succession might occur when a new habitat is opened up for colonization. In these cases, an entirely new community that is well adapted to the conditions found in this new area can establish itself.

River Continuum Concept

↑ Meandering stream in Waitomo, New Zealand

↑ River Gryffe in Scotland

↑ Rocky stream in Hawaii

The River continuum concept (RCC) was an attempt to construct a single framework to describe the function of temperate lotic ecosystems from the source to the end and relate it to changes in the biotic community (Vannote et al. 1980). The physical basis for RCC is size and location along the gradient from a small stream eventually linked to a large river. Stream order is used as the physical measure of the position along the RCC.

According to the RCC, low ordered sites are small shaded streams where allochthonous inputs of CPOM are a necessary resource for consumers. As the river widens at mid-ordered sites, energy inputs should change. Ample sunlight should reach the bottom in these systems to support significant periphyton production. Additionally, the biological processing of CPOM (Coarse Particulate Organic Matter - larger than 1 mm) inputs at upstream sites is expected to result in the transport of large amounts of FPOM (Fine Particulate Organic Matter - smaller than 1 mm) to these downstream ecosystems. Plants should become more abundant at edges of the river with increasing river size, especially in lowland rivers where finer sediments have been deposited and facilitate rooting. The main channels likely have too much current and turbidity and a lack of substrate to support plants or periphyton. Phytoplankton should produce the only autochthonous inputs here, but photosynthetic rates will be limited due to turbidity and mixing. Thus, allochthonous inputs are expected to be the primary energy source for large rivers. This FPOM will come from both upstream sites via the decomposition process and through lateral inputs from floodplains.

Biota should change with this change in energy from the headwaters to the mouth of these systems. Namely, shredders should prosper in low-ordered systems and grazers in mid-ordered sites. Microbial decomposition should play the largest role in energy production for low-ordered sites and large rivers, while photosynthesis, in addition to degraded allochthonous inputs from upstream will be essential in mid-ordered systems. As mid-ordered sites will theoretically receive the largest variety of energy inputs, they might be expected to host the most biological diversity (Vannote et al. 1980).

Just how well the RCC actually reflects patterns in natural systems is uncertain and its generality can be a handicap when applied to diverse and specific situations. The most noted criticisms of the RCC are: 1. It focuses mostly on macroinvertebrates, disregarding that plankton and fish diversity is highest in high orders; 2. It relies heavily on the fact that low ordered sites have high CPOM inputs, even though many streams lack riparian habitats; 3. It is based on pristine systems, which rarely exist today; and 4. It is centered around the functioning of temperate streams. Despite its shortcomings, the RCC remains a useful idea for describing how the patterns of ecological functions in a lotic system can vary from the source to the mouth.

Disturbances such as congestion by dams or natural events such as shore flooding are not included in the RCC model. Various researchers have since expanded the model to account for such irregularities. For example, J.V. Ward and J.A. Stanford came up with the Serial Discontinuity Concept in 1983, which addresses the impact of geomorphologic disorders such as congestion and integrated inflows. The same authors presented the Hyporheic Corridor concept in 1993, in which the vertical (in depth) and lateral (from shore to shore) structural complexity of the river were connected. The flood pulse concept, developed by W.J. Junk in 1989, further modified by P.B. Bayley in 1990 and K. Tockner in 2000, takes into account the large amount of nutrients and organic material that makes its way into a river from the sediment of surrounding flooded land.

Human Impacts

Pollution

Pollutant sources of lotic systems are hard to control because they derive, often in small amounts, over a very wide area and enter the system at many locations along its length. Agricultural fields often deliver large quantities of sediments, nutrients, and chemicals to nearby streams and rivers. Urban and residential areas can also add to this pollution when contaminants are accumulated on impervious surfaces such as roads and parking lots that then drain into the system. Elevated nutrient concentrations, especially nitrogen and phosphorus which are key components of fertilizers, can increase periphyton growth, which can be particularly dangerous in slow-moving streams. Another pollutant, acid rain, forms from sulfur dioxide and nitrous oxide emitted from factories and power stations. These substances readily dissolve in atmospheric moisture and enter lotic systems through precipitation. This can lower the pH of these sites, affecting all trophic levels from algae to vertebrates (Brown 1987). Mean species richness and total species numbers within a system decrease with decreasing pH.

While direct pollution of lotic systems has been greatly reduced in the United States under the government's Clean Water Act, contaminants from diffuse non-point sources remain a large problem.

Flow Modification

A weir on the River Calder, West Yorkshire

Dams alter the flow, temperature, and sediment regime of lotic systems. Additionally, many rivers are dammed at multiple locations, amplifying the impact. Dams can cause enhanced clarity and reduced variability in stream flow, which in turn cause an increase in periphyton abundance. Invertebrates immediately below a dam can show reductions in species richness due to an overall reduction in habitat heterogeneity. Also, thermal changes can affect insect development, with abnormally warm winter temperatures obscuring cues to break egg diapause and overly cool summer temperatures leaving too few acceptable days to complete growth. Finally, dams fragment river systems, isolating previously continuous populations, and preventing the migrations of anadromous and catadromous species.

Invasive Species

Invasive species have been introduced to lotic systems through both purposeful events (e.g. stocking game and food species) as well as unintentional events (e.g. hitchhikers on boats or fishing waders). These organisms can affect natives via competition for prey or habitat, predation, habitat alteration, hybridization, or the introduction of harmful diseases and parasites. Once established, these species can be difficult to control or eradicate, particularly because of the connectivity of lotic systems. Invasive species can be especially harmful in areas that have endangered biota, such as mussels in the Southeast United States, or those that have localized endemic species, like lotic systems west of the Rocky Mountains, where many species evolved in isolation.

References

- National Research Council (US) (1996) Freshwater ecosystems: revitalizing educational programs in limnology National Academy Press. ISBN 0-309-05443-5

- Brown, A. L. (1987). Freshwater Ecology. Heinimann Educational Books, London. p. 163. ISBN 0435606220.

- Brönmark, C.; L. A. Hansson (2005). The Biology of Lakes and Ponds. Oxford University Press, Oxford. p. 285. ISBN 0198516134.

- Giller, S.; B. Malmqvist (1998). The Biology of Streams and Rivers. Oxford University Press, Oxford. p. 296. ISBN 0198549776.

- Moss, B. (1998). Ecology of Freshwaters: man and medium, past to future. Blackwell Science, London. p. 557. ISBN 0632035129.

- Barange M, Field JG, Harris RP, Eileen E, Hofmann EE, Perry RI and Werner F (2010) Marine Ecosystems and Global Change Oxford University Press. ISBN 978-0-19-955802-5

- Boyd IL, Wanless S and Camphuysen CJ (2006) Top predators in marine ecosystems: their role in monitoring and management Volume 12 of Conservation biology series. Cambridge University Press. ISBN 978-0-521-84773-5

- Christensen V and Pauly D (eds.) (1993) Trophic models of aquatic ecosystems The WorldFish Center, issue 26 of ICLARM Technical Reports, volume 26 of ICLARM conference proceedings. ISBN 9789711022846.

- Davenport J (2008) Challenges to Marine Ecosystems: Proceedings of the 41st European Marine Biology Symposium Volume 202 of Developments in hydrobiology. ISBN 978-1-4020-8807-0

- Levner E, Linkov I and Proth J (2005) Strategic management of marine ecosystems Springer. Volume 50 of NATO Science Series IV. ISBN 978-1-4020-3158-8

- Mann KH and Lazier JRN (2006) Dynamics of marine ecosystems: biological-physical interactions in the oceans Wiley-Blackwell. ISBN 978-1-4051-1118-8

Marine Biology: An Integrated Approach

Marine biology is the study of all the organisms that are found in the ocean. There is a difference between marine biology and marine ecology; the difference being is that marine biology studies the interaction between organisms whereas marine ecology studies the interaction between the environment and the organisms. This section helps the readers in understanding marine biology in detail.

Marine Biology

Marine biology is the scientific study of organisms in the ocean or other marine bodies of water. Given that in biology many phyla, families and genera have some species that live in the sea and others that live on land, marine biology classifies species based on the environment rather than on taxonomy. Marine biology differs from marine ecology as marine ecology is focused on how organisms interact with each other and the environment, while biology is the study of the organisms themselves.

A large proportion of all life on Earth lives in the ocean. Exactly how large the proportion is unknown, since many ocean species are still to be discovered. The ocean is a complex three-dimensional world covering approximately 71% of the Earth's surface. The habitats studied in marine biology include everything from the tiny layers of surface water in which organisms and abiotic items may be trapped in surface tension between the ocean and atmosphere, to the depths of the oceanic trenches, sometimes 10,000 meters or more beneath the surface of the ocean. Specific habitats include coral reefs, kelp forests, seagrass meadows, the surrounds of seamounts and thermal vents, tidepools, muddy, sandy and rocky bottoms, and the open ocean (pelagic) zone, where solid objects are rare and the surface of the water is the only visible boundary. The organisms studied range from microscopic phytoplankton and zooplankton to huge cetaceans (whales) 30 meters (98 feet) in length.

Marine life is a vast resource, providing food, medicine, and raw materials, in addition to helping to support recreation and tourism all over the world. At a fundamental level, marine life helps determine the very nature of our planet. Marine organisms contribute significantly to the oxygen cycle, and are involved in the regulation of the Earth's climate. Shorelines are in part shaped and protected by marine life, and some marine organisms even help create new land.

Many species are economically important to humans, including both finfish and shell-fish. It is also becoming understood that the well-being of marine organisms and other organisms are linked in fundamental ways. The human body of knowledge regarding the relationship between life in the sea and important cycles is rapidly growing, with new discoveries being made nearly every day. These cycles include those of matter (such as the carbon cycle) and of air (such as Earth's respiration, and movement of energy through ecosystems including the ocean). Large areas beneath the ocean surface still remain effectively unexplored.

History

H.M.S. CHALLENGER UNDER SAIL, 1874.

HMS *Challenger* during its pioneer expedition of 1872–76

Early instances of the study of marine biology trace back to Aristotle (384–322 BC) who made several contributions which laid the foundation for many future discoveries and were the first big step in the early exploration period of the ocean and marine life. In 1768, Samuel Gottlieb Gmelin (1744–1774) published the *Historia Fucorum*, the first work dedicated to marine algae and the first book on marine biology to use the then new binomial nomenclature of Linnaeus. It included elaborate illustrations of seaweed and marine algae on folded leaves. The British naturalist Edward Forbes (1815–1854) is generally regarded as the founder of the science of marine biology. The pace of oceanographic and marine biology studies quickly accelerated during the course of the 19th century.

The observations made in the first studies of marine biology fueled the age of discovery and exploration that followed. During this time, a vast amount of knowledge was gained about the life that exists in the oceans of the world. Many voyages contributed significantly to this pool of knowledge. Among the most significant were the voyages of the HMS *Beagle* where Charles Darwin came up with his theories of evolution and on the formation of coral reefs. Another important expedition was undertaken by HMS *Challenger*, where findings were made of unexpectedly high species diversity among fauna stimulating much theorizing by population ecologists on how such varieties of life could be maintained in what was thought to be such a hostile environment. This era was important for the history of marine biology but naturalists were still limited in their studies because they lacked technology that would allow them to adequately examine species that lived in deep parts of the oceans.

The creation of marine laboratories was important because it allowed marine biologists to conduct research and process their specimens from expeditions. The oldest marine laboratory in the world, Station biologique de Roscoff, was established in France in 1872. In the United States, the Scripps Institution of Oceanography dates back to 1903, while the prominent Woods Hole Oceanographic Institute was founded in 1930. The development of technology such as sound navigation ranging, scuba diving gear, submersibles and remotely operated vehicles allowed marine biologists to discover and explore life in deep oceans that was once thought to not exist.

Marine Life

Microscopic Life

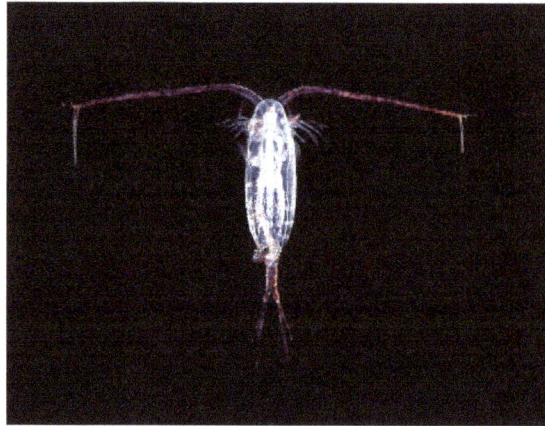

Copepod

As inhabitants of the largest environment on Earth, microbial marine systems drive changes in every global system. Microbes are responsible for virtually all the photosynthesis that occurs in the ocean, as well as the cycling of carbon, nitrogen, phosphorus and other nutrients and trace elements.

Microscopic life undersea is incredibly diverse and still poorly understood. For example, the role of viruses in marine ecosystems is barely being explored even in the beginning of the 21st century.

The role of phytoplankton is better understood due to their critical position as the most numerous primary producers on Earth. Phytoplankton are categorized into cyanobacteria (also called blue-green algae/bacteria), various types of algae (red, green, brown, and yellow-green), diatoms, dinoflagellates, euglenoids, coccolithophorids, cryptomonads, chrysophytes, chlorophytes, prasinophytes, and silicoflagellates.

Zooplankton tend to be somewhat larger, and not all are microscopic. Many Protozoa are zooplankton, including dinoflagellates, zooflagellates, foraminiferans, and radio-larians. Some of these (such as dinoflagellates) are also phytoplankton; the distinction between plants and animals often breaks down in very small organisms. Other zoo-plankton include cnidarians, ctenophores, chaetognaths, molluscs, arthropods, uro-chordates, and annelids such as polychaetes. Many larger animals begin their life as zooplankton before they become large enough to take their familiar forms. Two exam-ples are fish larvae and sea stars (also called starfish).

Plants and Algae

Microscopic algae and plants provide important habitats for life, sometimes acting as hiding and foraging places for larval forms of larger fish and invertebrates.

Algal life is widespread and very diverse under the ocean. Microscopic photosynthetic algae contribute a larger proportion of the world's photosynthetic output than all the terrestrial forests combined. Most of the niche occupied by sub plants on land is actual-ly occupied by macroscopic algae in the ocean, such as *Sargassum* and kelp, which are commonly known as seaweeds that create kelp forests.

Plants that survive in the sea are often found in shallow waters, such as the seagrasses (examples of which are eelgrass, *Zostera*, and turtle grass, *Thalassia*). These plants have adapted to the high salinity of the ocean environment. The intertidal zone is also a good place to find plant life in the sea, where mangroves or cordgrass or beach grass might grow. Microscopic algae and plants provide important habitats for life, sometimes acting as hiding and foraging places for larval forms of larger fish and invertebrates.

Invertebrates

Crown-of-thorns starfish

As on land, invertebrates make up a huge portion of all life in the sea. Invertebrate sea life includes Cnidaria such as jellyfish and sea anemones; Ctenophora; sea worms including the phyla Platyhelminthes, Nemertea, Annelida, Sipuncula, Echiura, Chaetognatha, and Phoronida; Mollusca including shellfish, squid, octopus; Arthropoda including Chelicerata and Crustacea; Porifera; Bryozoa; Echinodermata including starfish; and Urochordata including sea squirts or tunicates.

Fungi

Over 1500 species of fungi are known from marine environments. These are parasitic on marine algae or animals, or are saprobes on algae, corals, protozoan cysts, sea grasses, wood and other substrata, and can also be found in sea foam. Spores of many species have special appendages which facilitate attachment to the substratum. A very diverse range of unusual secondary metabolites is produced by marine fungi.

Vertebrates

Fish

Fish anatomy includes a two-chambered heart, operculum, swim bladder, scales, eyes adapted to seeing underwater, and secretory cells that produce mucous. Fish breathe by extracting oxygen from water through gills. Fins propel and stabilize the fish in the water. Fish fall into two main groups: fish with bony skeletons and fish with cartilaginous skeletons.

A reported 32,700 species of fish have been described (as of December 2013), more than the combined total of all other vertebrates. About 60% of fish species are saltwater fish.

Reptiles

Green turtle

Reptiles which inhabit or frequent the sea include sea turtles, sea snakes, terrapins, the marine iguana, and the saltwater crocodile. Most extant marine reptiles, except for some sea snakes, are oviparous and need to return to land to lay their eggs. Thus most species, excepting sea turtles, spend most of their lives on or near land rather than in the ocean. Despite their marine adaptations, most sea snakes prefer shallow waters nearby land, around islands, especially waters that are somewhat sheltered, as well as near estuaries. Some extinct marine reptiles, such as ichthyosaurs, evolved to be viviparous and had no requirement to return to land.

Birds

Birds adapted to living in the marine environment are often called seabirds. Examples include albatross, penguins, gannets, and auks. Although they spend most of their lives in the ocean, species such as gulls can often be found thousands of miles inland.

Mammals

Sea otters

There are five main types of marine mammals.

- Cetaceans include toothed whales (suborder Odontoceti), such as the sperm whale, dolphins, and porpoises such as the Dall's porpoise. Cetaceans also include baleen whales (suborder Mysticeti), such as the gray whale, humpback whale, and blue whale.

- Sirenians include manatees, the dugong, and the extinct Steller's sea cow.

- Seals (family Phocidae), sea lions (family Otariidae - which also include the fur seals), and the walrus (family Odobenidae) are all considered pinnipeds.

- The sea otter is a member of the family Mustelidae, which includes weasels and badgers.

- The polar bear is a member of the family Ursidae.

Marine Habitats

Marine habitats can be divided into coastal and open ocean habitats. Coastal habitats are found in the area that extends from the shoreline to the edge of the continental shelf. Most marine life is found in coastal habitats, even though the shelf area occupies only seven percent of the total ocean area. Open ocean habitats are found in the deep ocean beyond the edge of the continental shelf

Alternatively, marine habitats can be divided into pelagic and demersal habitats. Pelagic habitats are found near the surface or in the open water column, away from the bottom of the ocean. Demersal habitats are near or on the bottom of the ocean. An organism living in a pelagic habitat is said to be a pelagic organism, as in pelagic fish. Similarly, an organism living in a demersal habitat is said to be a demersal organism, as in demersal fish. Pelagic habitats are intrinsically shifting and ephemeral, depending on what ocean currents are doing.

Marine habitats can be modified by their inhabitants. Some marine organisms, like corals, kelp and seagrasses, are ecosystem engineers which reshape the marine environment to the point where they create further habitat for other organisms.

Intertidal and Near Shore

Tide pools with sea stars and sea anemone

Intertidal zones, those areas close to shore, are constantly being exposed and covered by the ocean's tides. A huge array of life lives within this zone.

Shore habitats span from the upper intertidal zones to the area where land vegetation takes prominence. It can be underwater anywhere from daily to very infrequently. Many species here are scavengers, living off of sea life that is washed up on the shore. Many land animals also make much use of the shore and intertidal habitats. A subgroup of organisms in this habitat bores and grinds exposed rock through the process of bioerosion.

Estuaries

Estuaries have shifting flows of sea water and fresh water.

Estuaries are also near shore and influenced by the tides. An estuary is a partially en-closed coastal body of water with one or more rivers or streams flowing into it and with a free connection to the open sea. Estuaries form a transition zone between freshwa-ter river environments and saltwater maritime environments. They are subject both to marine influences—such as tides, waves, and the influx of saline water—and to riverine influences—such as flows of fresh water and sediment. The shifting flows of both sea water and fresh water provide high levels of nutrients both in the water column and in sediment, making estuaries among the most productive natural habitats in the world.

Reefs

Coral reefs form complex marine ecosystems with tremendous biodiversity.

Reefs comprise some of the densest and most diverse habitats in the world. The best-known types of reefs are tropical coral reefs which exist in most tropical wa-ters; however, reefs can also exist in cold water. Reefs are built up by corals and other calcium-depositing animals, usually on top of a rocky outcrop on the ocean floor. Reefs can also grow on other surfaces, which has made it possible to create artificial reefs. Coral reefs also support a huge community of life, including the corals themselves, their symbiotic zooxanthellae, tropical fish and many other or-ganisms.

Much attention in marine biology is focused on coral reefs and the El Niño weather phenomenon. In 1998, coral reefs experienced the most severe mass bleaching events on record, when vast expanses of reefs across the world died because sea surface tem-peratures rose well above normal. Some reefs are recovering, but scientists say that between 50% and 70% of the world's coral reefs are now endangered and predict that global warming could exacerbate this trend.

Open Ocean

The open ocean is relatively unproductive because of a lack of nutrients, yet because it is so vast, in total it produces the most primary productivity. Much of the aphotic zone's energy is supplied by the open ocean in the form of detritus.

The open ocean is the area of deep sea beyond the continental shelves

Deep Sea and Trenches

The deepest recorded oceanic trench measured to date is the Mariana Trench, near the Philippines, in the Pacific Ocean at 10,924 m (35,840 ft). At such depths, water pressure is extreme and there is no sunlight, but some life still exists. A white flatfish, a shrimp and a jellyfish were seen by the American crew of the bathyscaphe *Trieste* when it dove to the bottom in 1960.

Other notable oceanic trenches include Monterey Canyon, in the eastern Pacific, the Tonga Trench in the southwest at 10,882 m (35,702 ft), the Philippine Trench, the Puerto Rico Trench at 8,605 m (28,232 ft), the Romanche Trench at 7,760 m (25,460 ft), Fram Basin in the Arctic Ocean at 4,665 m (15,305 ft), the Java Trench at 7,450 m (24,440 ft), and the South Sandwich Trench at 7,235 m (23,737 ft).

In general, the deep sea is considered to start at the aphotic zone, the point where sunlight loses its power of transference through the water. Many life forms that live at these depths have the ability to create their own light known as bio-luminescence.

Marine life also flourishes around seamounts that rise from the depths, where fish and other sea life congregate to spawn and feed. Hydrothermal vents along the mid-ocean ridge spreading centers act as oases, as do their opposites, cold seeps. Such places support unique biomes and many new microbes and other lifeforms have been discovered at these locations .

Subfields

The marine ecosystem is large, and thus there are many sub-fields of marine biology. Most involve studying specializations of particular animal groups, such as phycology, invertebrate zoology and ichthyology.

Other subfields study the physical effects of continual immersion in sea water and the ocean in general, adaptation to a salty environment, and the effects of changing various oceanic properties on marine life. A subfield of marine biology studies the relationships between oceans and ocean life, and global warming and environmental issues (such as carbon dioxide displacement).

Recent marine biotechnology has focused largely on marine biomolecules, especially proteins, that may have uses in medicine or engineering. Marine environments are the home to many exotic biological materials that may inspire biomimetic materials.

Related Fields

Marine biology is a branch of biology. It is closely linked to oceanography and may be regarded as a sub-field of marine science. It also encompasses many ideas from ecology. Fisheries science and marine conservation can be considered partial offshoots of marine biology (as well as environmental studies). Marine Chemistry, Physical oceanography and Atmospheric sciences are closely related to this field.

Distribution Factors

An active research topic in marine biology is to discover and map the life cycles of various species and where they spend their time. Technologies that aid in this discovery include pop-up satellite archival tags, acoustic tags, and a variety of other data loggers. Marine biologists study how the ocean currents, tides and many other oceanic factors affect ocean life forms, including their growth, distribution and well-being. This has only recently become technically feasible with advances in GPS and newer underwater visual devices.

Most ocean life breeds in specific places, nests or not in others, spends time as juveniles in still others, and in maturity in yet others. Scientists know little about where many species spend different parts of their life cycles especially in the infant and juvenile years. For example, it is still largely unknown where juvenile sea turtles and some year-1 sharks travel. Recent advances in underwater tracking devices are illuminating what we know about marine organisms that live at great Ocean depths. The information that pop-up satellite archival tags give aids in certain time of the year fishing closures and development of a marine protected area. This data is important to both scientists and fishermen because they are discovering that by restricting commercial fishing in one small area they can have a large impact in maintaining a healthy fish population in a much larger area.

Marine Invertebrates

Marine invertebrates are the invertebrates that live in marine habitats. Invertebrate is a blanket term that includes all animals apart from the vertebrate members of the chordate phylum. Invertebrates lack a vertebral column, and some have evolved a shell or a hard exoskeleton. As on land and in the air, marine invertebrates have a large variety of body plans, and have been categorised into over 30 phyla. They make up most of the macroscopic life in the oceans.

Evolution

Kimberella, an early mollusc important for understanding the Cambrian explosion. Invertebrates are grouped into different phyla (body plans).

Opabinia an extinct, stem group arthropod that appeared in the Middle Cambrian

The earliest animals were marine invertebrates, that is, vertebrates came later. Animals are multicellular eukaryotes, and are distinguished from plants, algae, and fungi by lacking cell walls. Marine invertebrates are animals that inhabit a marine environment apart from the vertebrate members of the chordate phylum; invertebrates lack a vertebral column. Some have evolved a shell or a hard exoskeleton.

The earliest widely accepted animal fossils are the rather modern-looking cnidarians (the group that includes jellyfish, sea anemones and *Hydra*), possibly from around 580 Ma The Ediacara biota, which flourished for the last 40 million years before the start of the Cambrian, were the first animals more than a very few centimetres long. Many were flat and had a "quilted" appearance, and seemed so strange that there was a proposal to classify them as a separate kingdom, Vendozoa. Others, however, have been interpreted as early molluscs (*Kimberella*), echinoderms (*Arkarua*), and arthropods (*Spriggina, Parvancorina*). There is still debate about the classification of these specimens, mainly because the diagnostic features which allow taxonomists to classify more recent organisms, such as similarities to living organisms, are generally absent in the Ediacarans. However, there seems little doubt that *Kimberella* was at least a triploblastic bilaterian animal, in other words, an animal significantly more complex than the cnidarians.

The small shelly fauna are a very mixed collection of fossils found between the Late Ediacaran and Middle Cambrian periods. The earliest, *Cloudina*, shows signs of successful defense against predation and may indicate the start of an evolutionary arms race. Some tiny Early Cambrian shells almost certainly belonged to molluscs, while the owners of some "armor plates," *Halkieria* and *Microdictyon*, were eventually identified when more complete specimens were found in Cambrian lagerstätten that preserved soft-bodied animals.

In the 1970s there was already a debate about whether the emergence of the modern phyla was "explosive" or gradual but hidden by the shortage of Precambrian animal fossils. A re-analysis of fossils from the Burgess Shale lagerstätte increased interest in the issue when it revealed animals, such as *Opabinia*, which did not fit into any known phylum. At the time these were interpreted as evidence that the modern phyla had evolved very rapidly in the Cambrian explosion and that the Burgess Shale's "weird wonders" showed that the Early Cambrian was a uniquely experimental period of animal evolution. Later discoveries of similar animals and the development of new theoretical approaches led to the conclusion that many of the "weird wonders" were evolutionary "aunts" or "cousins" of modern groups—for example that *Opabinia* was a member of the lobopods, a group which includes the ancestors of the arthropods, and that it may have been closely related to the modern tardigrades. Nevertheless, there is still much debate about whether the Cambrian explosion was really explosive and, if so, how and why it happened and why it appears unique in the history of animals.

Classification

Invertebrates are grouped into different phyla. Informally phyla can be thought of as a way of grouping organisms according to their body plan. A body plan refers to a blueprint which describes the shape or morphology of an organism, such as its symmetry, segmentation and the disposition of its appendages. The idea of body plans originated with vertebrates, which were grouped into one phylum. But the vertebrate body plan is only one of many, and invertebrates consist of many phyla or body plans. The history of the discovery of body plans can be seen as a movement from a worldview centred on vertebrates, to seeing the vertebrates as one body plan among many. Among the pioneering zoologists, Linnaeus identified two body plans outside the vertebrates; Cuvier identified three; and Haeckel had four, as well as the Protista with eight more, for a total of twelve. For comparison, the number of phyla recognised by modern zoologists has risen to 35.

Bryozoa, from Ernst Haeckel's *Kunstformen der Natur*, 1904

Historically body plans were thought of as having evolved in rapidly during the Cambrian explosion, but a more nuanced understanding of animal evolution suggests a gradual development of body plans throughout the early Palaeozoic and beyond. More generally a phylum can be defined in two ways: as described above, as a group of organisms with a certain degree of morphological or developmental similarity (the phenetic definition), or a group of organisms with a certain degree of evolutionary relatedness (the phylogenetic definition).

As on land and in the air, invertebrates make up a great majority of all macroscopic life, as the vertebrates makes up a subphylum of one of over 30 known animal phyla, making the term almost meaningless for taxonomic purpose. Invertebrate sea life includes the following groups, some of which are phyla:

The 49th plate from Ernst Haeckel's *Kunstformen der Natur*, 1904, showing various sea anemones classified as Actiniae, in the Cnidaria phylum

"A variety of marine worms": plate from *Das Meer* by M.J. Schleiden (1804–1881)

- Acoela, among the most primitive bilateral animals;

- Annelida, (polychaetes and sea leeches);

- Brachiopoda, marine animals that have hard "valves" (shells) on the upper and lower surfaces ;

- Bryozoa, also known as moss animals or sea mats;

- Chaetognatha, commonly known as arrow worms, are a phylum of predatory marine worms that are a major component of plankton;

- Cephalochordata represented in the modern oceans by the lancelets (also known as Amphioxus);

- Cnidaria, such as jellyfish, sea anemones, and corals;

- Crustacea, including lobsters, crabs, shrimp, crayfish, barnacles, hermit crabs, mantis shrimps, and copepods;

- Ctenophora, also known as comb jellies, the largest animals that swim by means of cilia;

- Echinodermata, including sea stars, brittle stars, sea urchins, sand dollars, sea cucumbers, crinoids, and sea daisies;

- Echiura, also known as spoon worms;

- Gnathostomulids, slender to thread-like worms, with a transparent body that inhabit sand and mud beneath shallow coastal waters;

- Gastrotricha, often called hairy backs, found mostly interstitially in between sediment particles;

- Hemichordata, includes acorn worms, solitary worm-shaped organisms;

- Kamptozoa, goblet-shaped sessile aquatic animals, with relatively long stalks and a "crown" of solid tentacles, also called Entoprocta;

- Kinorhyncha, segmented, limbless animals, widespread in mud or sand at all depths, also called mud dragons;

- Loricifera, very small to microscopic marine sediment-dwelling animals only discovered in 1983;

- Merostomata; also known as horseshoe crabs;

- Mollusca, including shellfish, squid, octopus, whelks, *Nautilus*, cuttlefish, nudibranchs, scallops, sea snails, Aplacophora, Caudofoveata, Monoplacophora, Polyplacophora, and Scaphopoda;

- Myzostomida, a taxonomic group of small marine worms which are parasitic on crinoids or "sea lilies";

- Nemertinea, also known as "ribbon worms" or "proboscis worms";

- Orthonectida, a small phylum of poorly known parasites of marine invertebrates that are among the simplest of multi-cellular organisms;

- Phoronida, a phylum of marine animals that filter-feed with a lophophore (a "crown" of tentacles), and build upright tubes of chitin to support and protect their soft bodies;

- Placozoa, small, flattened, multicellular animals around 1 millimetre across and the simplest in structure. They have no regular outline, although the lower surface is somewhat concave, and the upper surface is always flattened;

- Porifera (sponges), multicellular organisms that have bodies full of pores and channels allowing water to circulate through them;

- Priapulida, or penis worms, are a phylum of marine worms that live marine mud. They are named for their extensible spiny proboscis, which, in some species, may have a shape like that of a human penis;

- Pycnogonida, also called sea spiders, are unrelated to spiders, or even to arachnids which they resemble;

- Sipunculida, also called peanut worms, is a group containing 144–320 species (estimates vary) of bilaterally symmetrical, unsegmented marine worms;

- Tunicata, also known as sea squirts or sea pork, are filter feeders attached to rocks or similarly suitable surfaces on the ocean floor;

- Some flatworms of the classes Turbellaria and Monogenea;

- Xenoturbella, a genus of bilaterian animals that contains only two marine worm-like species;

- Xiphosura, includes a large number of extinct lineages and only four recent species in the family Limulidae, which include the horseshoe crabs.

Arthropods total about 1,113,000 described extant species, molluscs about 85,000 and chordates about 52,000.

Marine Sponges

Sponges are animals of the phylum Porifera (Modern Latin for *bearing pores*). They are multicellular organisms that have bodies full of pores and channels allowing water to circulate through them, consisting of jelly-like mesohyl sandwiched between two thin layers of cells. They have unspecialized cells that can transform into other types and that often migrate between the main cell layers and the mesohyl in the process. Sponges do not have nervous, digestive or circulatory systems. Instead, most rely on maintaining a constant water flow through their bodies to obtain food and oxygen and to remove wastes.

Sponges are similar to other animals in that they are multicellular, heterotrophic, lack cell walls and produce sperm cells. Unlike other animals, they lack true tissues and organs, and have no body symmetry. The shapes of their bodies are adapted for maximal efficiency of water flow through the central cavity, where it deposits nutrients, and leaves through a hole called the osculum. Many sponges have internal skeletons of spongin and/or spicules of calcium carbonate or silicon dioxide. All sponges are sessile aquatic animals. Although there are freshwater species, the great majority are marine (salt water) species, ranging from tidal zones to depths exceeding 8,800 m (5.5 mi).

Sponge biodiversity. There are four sponge species in this photo.

While most of the approximately 5,000–10,000 known species feed on bacteria and other food particles in the water, some host photosynthesizing micro-organisms as endosymbionts and these alliances often produce more food and oxygen than they consume. A few species of sponge that live in food-poor environments have become carnivores that prey mainly on small crustaceans.

Branching vase sponge

Venus' flower basket at a depth of 2572 meters

Barrel sponge

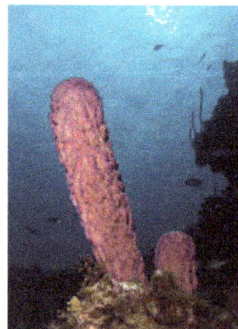

Stove-pipe sponge

Linnaeus mistakenly identified sponges as plants in the order Algae. For a long time thereafter sponges were assigned to a separate subkingdom, Parazoa (meaning *beside the animals*). They are now classified as a paraphyletic phylum from which the higher animals have evolved.

Marine Cnidarians

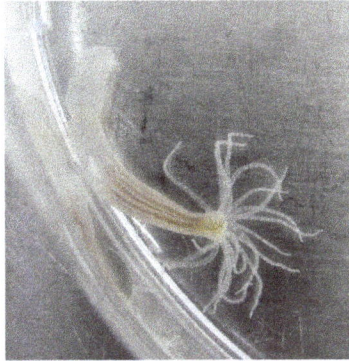

Cnidarians are the simplest animals with cells organised into tissues. Yet the starlet sea anemone contains the same genes as those that form the vertebrate head.

Cnidarians (Greek for *nettle*) are distinguished by the presence of stinging cells, specialized cells that they use mainly for capturing prey. Cnidarians include corals, sea anemones, jellyfish and hydrozoans. They form a phylum containing over 10,000 species of animals found exclusively in aquatic (mainly marine) environments. Their bodies consist of mesoglea, a non-living jelly-like substance, sandwiched between two layers of epithelium that are mostly one cell thick. They have two basic body forms: swimming medusae and sessile polyps, both of which are radially symmetrical with mouths surrounded by tentacles that bear cnidocytes. Both forms have a single orifice and body cavity that are used for digestion and respiration.

Fossil cnidarians have been found in rocks formed about 580 million years ago. Fossils of cnidarians that do not build mineralized structures are rare. Scientists currently think cnidarians, ctenophores and bilaterians are more closely related to calcareous sponges than these are to other sponges, and that anthozoans are the evolutionary "aunts" or "sisters" of other cnidarians, and the most closely related to bilaterians.

Cnidarians are the simplest animals in which the cells are organised into tissues. The starlet sea anemone is used as a model organism in research. It is easy to care for in the laboratory and a protocol has been developed which can yield large numbers of embryos on a daily basis. There is a remarkable degree of similarity in the gene sequence conservation and complexity between the sea anemone and vertebrates. In particular, genes concerned in the formation of the head in vertebrates are also present in the anemone.

Sea anemones are common in tidepools

Their tentacles sting and paralyse small fish

Close up of polyps on the surface of a coral, waving their tentacles.

Marine Worms

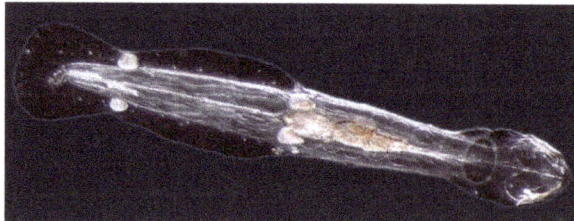
Arrow worms are predatory components of plankton worldwide.

Worms (Old English for *serpent*) typically have long cylindrical tube-like bodies and no limbs. Marine worms vary in size from microscopic to over 1 metre (3.3 ft) in length for some marine polychaete worms (bristle worms) and up to 58 metres (190 ft) for the marine nemertean worm (bootlace worm). Some marine worms occupy a small variety of parasitic niches, living inside the bodies of other animals, while others live more freely in the marine environment or by burrowing underground.

Different groups of marine worms are related only distantly, so they are found in several different phyla such as the Annelida (segmented worms), Chaetognatha (arrow worms), Hemichordata, and Phoronida (horseshoe worms). Many of these worms have specialized tentacles used for exchanging oxygen and carbon dioxide and also may be used for reproduction. Some marine worms are tube worms, such as the giant tube worm which lives in waters near underwater volcanoes and can withstand temperatures up to 90 degrees Celsius.

Platyhelminthes (flatworms) form another worm phylum which includes a class Cestoda of parasitic tapeworms. The marine tapeworm *Polygonoporus giganticus*, found in the gut of sperm whales, can grow to over 30 m (100 ft).

The bootlace worm can grow to 58 metres (190 ft).

Nematodes (roundworms) constitute a further worm phylum with tubular digestive systems and an opening at both ends. Over 25,000 nematode species have been described, of which more than half are parasitic, t, and it has been estimated another million remain undescribed. They are ubiquitous in marine, freshwater and terrestrial environments, where they often outnumber other animals in both individual and species counts. They are found in every part of the earth's lithosphere, from the top of mountains to the bottom of oceanic trenches. By count they represent 90% of all animals on the ocean floor. Their numerical dominance, often exceeding a million individuals per square meter and accounting for about 80% of all individual animals on earth, their diversity of life cycles, and their presence at various trophic levels point at an important role in many ecosystems.

Giant tube worms cluster around hydrothermal vents

Lamellibrachia luymes, a cold seep tubeworm, lives over 250 years.

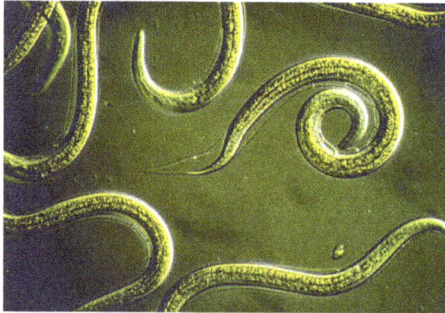
Nematodes are ubiquitous pseudocoelomates which can parasite marine plants and animals.

Bloodworms are typically found on the bottom of shallow marine waters

Echinoderms

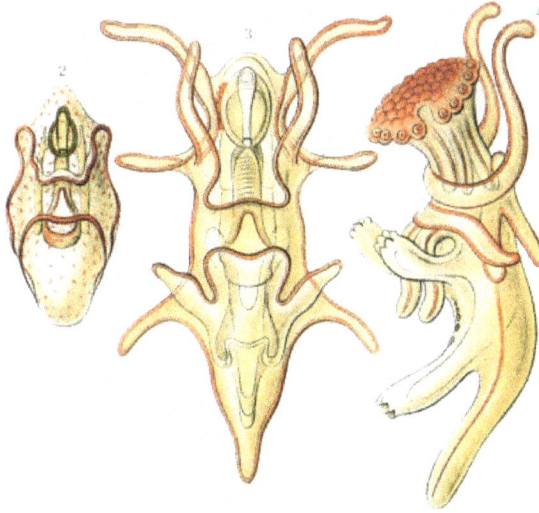

Starfish larvae are bilaterally symmetric, whereas the adults have fivefold symmetry

Echinoderms (Greek for *spiny skin*) is a phylum which contains only marine inverte-brates. The adults are recognizable by their radial symmetry (usually five-point) and include starfish, sea urchins, sand dollars, and sea cucumbers, as well as the sea lilies. Echinoderms are found at every ocean depth, from the intertidal zone to the abyssal zone. The phylum contains about 7000 living species, making it the second-largest grouping of deuterostomes (a superphylum), after the chordates (which include the vertebrates, such as birds, fishes, mammals, and reptiles).

Echinoderms are unique among animals in having bilateral symmetry at the larval stage, but fivefold symmetry (pentamerism, a special type of radial symmetry) as adults.

The echinoderms are important both biologically and geologically. Biologically, there are few other groupings so abundant in the biotic desert of the deep sea, as well as shallower oceans. The more notably distinct trait, which most echino-derms have, is their remarkable powers of regeneration of tissue, organs, limbs, and of asexual reproduction, and in some cases, complete regeneration from a single limb. Geologically, the value of echinoderms is in their ossified skeletons, which are major contributors to many limestone formations, and can provide valuable clues as to the geological environment. They were the most used species in regen-erative research in the 19th and 20th centuries. Further, it is held by some scien-tists that the radiation of echinoderms was responsible for the Mesozoic Marine Revolution.

Aside from the hard-to-classify *Arkarua* (a Precambrian animal with echinoderm-like pentamerous radial symmetry), the first definitive members of the phylum appeared near the start of the Cambrian.

Echinoderm literally means "spiny skin", as this water melon sea urchin illustrates

Sea cucumbers filter feed on plankton and suspended solids

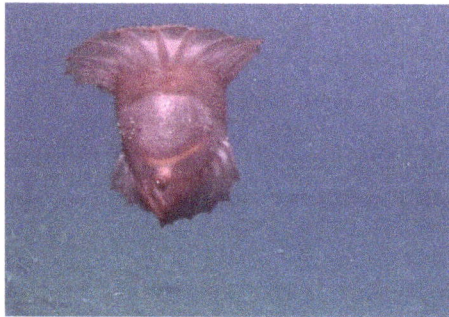

Benthopelagic sea cucumbers can lift off the seafloor and journey as much as 1,000 m (3,300 ft) up the water column

The ochre sea star was the first keystone predator to be studied. They limit mussels which can overwhelm intertidal communities.

Colorful sea lilies in shallow waters

Marine Molluscs

Molluscs (Latin for *soft*) form a phylum with about 85,000 extant recognized species. They are the largest marine phylum, comprising about 23% of all the named marine organisms. Molluscs have more varied forms than other invertebrate phylums. They are highly diverse, not just in size and in anatomical structure, but also in behaviour and in habitat. The majority of species still live in the oceans, from the seashores to the abyssal zone, but some form a significant part of the freshwater fauna and the terrestrial ecosystems.

Reconstruction of an ammonite, a highly successful early cephalopod that first appeared in the Devonian (about 400 mya). They became extinct during the same extinction event that killed the land dinosaurs (about 66 mya).

The mollusc phylum is divided into 9 or 10 taxonomic classes, two of which are extinct. These classes include gastropods, bivalves and cephalopods, as well as other lesser-known but distinctive classes. Gastropods with protective shells are referred to as snails, whereas gastropods without protective shells are referred to as slugs. Gastropods are by far the most numerous molluscs in terms of classified species, accounting for 80% of the total. Bivalves include clams, oysters, cockles, mussels, scallops, and numerous other families. There are about 8,000 marine bivalves species (including brackish water and estuarine species), and about 1,200 freshwater species. Cephalopod include octopus, squid and cuttlefish. They are found in all oceans, and neurologically are the most advanced of the invertebrates. About 800 living species of marine cephalopods have been identified, and an estimated 11,000 extinct taxa have been described. There are no fully freshwater cephalopods.

Colossal squid, largest of all invertebrates

The nautilus is a living fossil little changed since it evolved 500 million years ago as one of the first cephalopods.

Molluscs have such diverse shapes that many textbooks base their descriptions of molluscan anatomy on a generalized or *hypothetical ancestral mollusc*. This generalized mollusc is unsegmented and bilaterally symmetrical with an underside consisting of a single muscular foot. Beyond that it has three further key features. Firstly, it has a muscular cloak called a mantle covering its viscera and containing a significant cavity used for breathing and excretion. A shell secreted by the mantle covers the upper surface. Secondly (apart from bivalves) it has a rasping tongue called a radula used for feeding. Thirdly, it has a nervous system including a complex digestive system using microscopic, muscle-powered hairs called cilia to exude mucus. The generalized mollusc has two paired nerve cords (three in bivalves). The brain, in species that have one, encircles the esophagus. Most molluscs have eyes and all have sensors detecting chemicals, vibrations, and touch. The simplest type of molluscan reproductive system relies on external fertilization, but more complex variations occur. All produce eggs, from which may emerge trochophore larvae, more complex veliger larvae, or miniature adults. The depiction is rather similar to modern monoplacophorans, and some suggest it may resemble very early molluscs.

Marine gastropods are sea snails or sea slugs. This nudibranch is a sea slug.

Generalized or *hypothetical ancestral mollusc*

Good evidence exists for the appearance of marine gastropods, cephalopods and bivalves in the Cambrian period 541 to 485.4 million years ago. However, the evolutionary history both of molluscs' emergence from the ancestral Lophotrochozoa and of their diversification into the well-known living and fossil forms are still subjects of vigorous debate among scientists.

Marine Arthropods

Arthropods (Greek for *jointed feet*) have an exoskeleton (external skeleton), a segmented body, and jointed appendages (paired appendages). They form a phylum which includes insects, arachnids, myriapods, and crustaceans. Arthropods are characterized by their jointed limbs and cuticle made of chitin, often mineralised with calcium carbonate. The arthropod body plan consists of segments, each with a pair of appendages. The rigid cuticle inhibits growth, so arthropods replace it periodically by moulting. Their versatility has enabled them to become the most species-rich members of all ecological guilds in most environments.

Head

Thorax

Abdomen

Segmentation and tagmata of an arthropod

Marine arthropods range in size from the microscopic crustacean *Stygotantulus* to the Japanese spider crab. Arthropods' primary internal cavity is a hemocoel, which accommodates their internal organs, and through which their haemolymph - analogue of blood - circulates; they have open circulatory systems. Like their exteriors, the internal organs of arthropods are generally built of repeated segments. Their nervous system is "ladder-like", with paired ventral nerve cords running through all segments and forming paired ganglia in each segment. Their heads are formed by fusion of varying numbers of segments, and their brains are formed by fusion of the ganglia of these segments and encircle the esophagus. The respiratory and excretory systems of arthropods vary, depending as much on their environment as on the subphylum to which they belong.

Their vision relies on various combinations of compound eyes and pigment-pit ocelli: in most species the ocelli can only detect the direction from which light is coming, and the compound eyes are the main source of information, but the main eyes of spiders are ocelli that can form images and, in a few cases, can swivel to track prey. Arthropods also have a wide range of chemical and mechanical sensors, mostly based on modifications of the many setae (bristles) that project through their cuticles. Arthropods' methods of reproduction and development are diverse; all terrestrial species use internal fertilization, but this is often by indirect transfer of the sperm via an appendage or the ground, rather than by direct injection. Marine species all lay eggs and use either internal or external fertilization. Arthropod hatchlings vary from miniature adults to grubs that lack jointed limbs and eventually undergo a total metamorphosis to produce the adult form.

Trilobites, now extinct, roamed oceans for 270 million years.

Horseshoe crab, a living fossil arthropod from 450 million years ago

- Crustaceans

Many crustaceans are very small, like this tiny amphipod, and make up a significant part of the ocean's zooplankton

The evolutionary ancestry of arthropods dates back to the Cambrian period. The group is generally regarded as monophyletic, and many analyses support the placement of arthropods with cycloneuralians (or their constituent clades) in a superphylum Ecdysozoa. Overall however, the basal relationships of Metazoa are not yet well resolved. Likewise, the relationships between various arthropod groups are still actively debated.

Other Phyla

- Tardigrade, Lobopodia, (Onychophora)

- Non-craniate (non-vertebrate) chordates: Cephalochordate, Tunicata and *Haikouella*. These invertebrates are close relatives of the vertebrates.

- Non-craniate chordates are close relatives of vertebrates

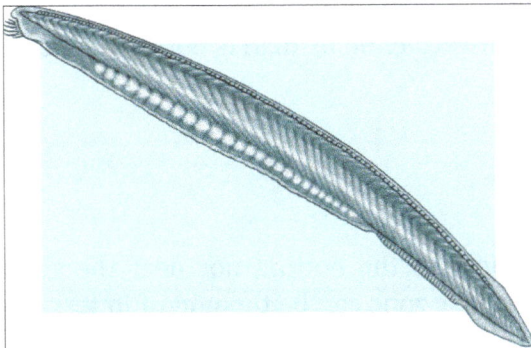

The lancelet, a small translucent fish-like Cephalochordate, is the closest living invertebrate relative of the vertebrates.

Fluorescent-colored sea squirts, *Rhopalaea crassa*. Tunicates may provide clues to vertebrate (and therefore human) ancestry.

Salp chain

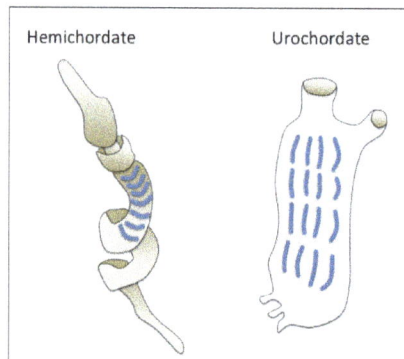

Gill slits in an acorn worm (left) and tunicate (right)

Minerals From Sea Water

There are a number of marine invertebrates that use minerals that are present in the sea in such minute quantities that they were undetectable until the advent of spectroscopy. Vanadium is concentrated by some tunicates for use in their blood cells to a level ten million times that of the surrounding seawater. Other tunicates similarly concentrate niobium and tantalum. Lobsters use copper in their respiratory pigment hemocyanin, despite the proportion of this metal in seawater being minute. Although these elements are present in vast quantities in the ocean, their extraction by man is not economic.

Pelagic Zone

Any water in a sea or lake that is neither close to the bottom nor near the shore can be said to be in the pelagic zone. The pelagic zone can be thought of in terms of an imaginary cylinder or water column that goes from the surface of the sea almost to the bottom. Conditions differ deeper in the water column such that as pressure increases with depth, the temperature drops and less light penetrates. Depending on the depth, the water column, rather like the Earth's atmosphere, may be divided into different layers.

The pelagic zone occupies 1,330 million km³ (320 million mi³) with a mean depth of 3.68 km (2.29 mi) and maximum depth of 11 km (6.8 mi). Fish that live in the pelagic zone are called pelagic fish. Pelagic life decreases with increasing depth. It is affected by light intensity, pressure, temperature, salinity, the supply of dissolved oxygen and nutrients, and the submarine topography, which is called bathymetry. In deep water, the pelagic zone is sometimes called the open-ocean zone and can be contrasted with water that is near the coast or on the continental shelf. In other contexts, coastal water not near the bottom is still said to be in the pelagic zone.

The pelagic zone can be contrasted with the benthic and demersal zones at the bottom of the sea. The benthic zone is the ecological region at the very bottom of the sea. It includes the sediment surface and some subsurface layers. Marine organisms living in this zone, such as clams and crabs, are called benthos. The demersal zone is just above the benthic zone. It can be significantly affected by the seabed and the life that lives there. Fish that live in the demersal zone are called demersal fish, which can be divided into benthic fish, which are denser than water so they can rest on the bottom, and benthopelagic fish, which swim in the water column just above the bottom. Demersal fish are also known as bottom feeders and groundfish.

Depth and Layers

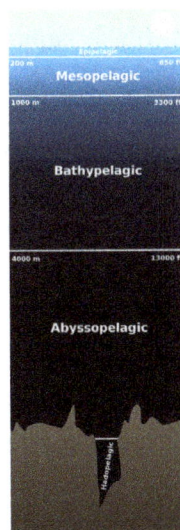

A scale diagram of the layers of the pelagic zone

Depending on how deep the sea is, the pelagic zone can extend over up to five horizontal layers in the ocean. From the top down, these are:

Epipelagic (Sunlight)

From the surface (MSL) down to around 200 m (660 ft)

This is the illuminated zone at the surface of the sea where enough light is available for photosynthesis. Nearly all primary production in the ocean occurs here. Consequently, plants and animals are largely concentrated in this zone. Examples of organisms living in this zone are plankton, floating seaweed, jellyfish, tuna, many sharks and dolphins.

Mesopelagic (Twilight)

From 200 m (660 ft) down to around 1,000 m (3,300 ft)

The name for this zone stems from Greek μέσον *(meson)*, meaning "middle". Although some light penetrates this second layer it is insufficient for photosynthesis. At about 500 m the water also becomes depleted of oxygen. Organisms survive this environment by having more efficient gills or by minimizing movement.

Examples of animals that live here are swordfish, squid, wolffish and some species of cuttlefish. Many organisms that live in this zone are bioluminescent. Some creatures living in the mesopelagic zone rise to the epipelagic zone at night to feed.

Bathypelagic (Midnight)

From 1,000 m (3,300 ft) down to around 4,000 m (13,000 ft)

At this depth, the ocean is pitch black, apart from occasional bioluminescent organisms, such as lanternfish. No living plant life exists here. Most animals living here survive by consuming the detritus falling from the zones above, which is known as "marine snow", or, like the marine hatchetfish, by preying on other inhabitants of this zone. Other examples of this zone's inhabitants are giant squid, smaller squids and the dumbo octopus. The giant squid is hunted here by deep-diving sperm whales.

Abyssopelagic (Lower Midnight)

From around 4,000 m (13,000 ft) down to above the ocean floor

Very few creatures live in the cold temperatures, high pressures and complete darkness of this depth. Among the species found in this zone are several species of squid; echinoderms including the basket star, swimming cucumber, and the sea pig; and marine arthropods including the sea spider. Many of the species living at these depths are transparent and eyeless because of the total lack of light in this zone.

Hadopelagic

The deep water in ocean trenches

The name is derived from the realm of Hades, the underworld in Greek mythology.

This zone is mostly unknown, and very few species are known to live here (in the open areas). However, many organisms live in hydrothermal vents in this and other zones. Some define the hadopelagic as waters below 6,000 m (20,000 ft), whether in a trench or not. The bathypelagic, abyssopelagic, and hadopelagic zones are very similar in character, and some marine biologists combine them into a single zone or consider the latter two to be the same. The abyssal plain is covered with soft sludge composed of dead organisms from above.

Pelagic Ecosystem

The pelagic sooty tern spends months at a time flying at sea, returning to land only for breeding.

The pelagic ecosystem is based on the phytoplankton which occupy the start of the foodchain. Phytoplankton manufacture their own food using a process of photosynthesis. Because they need sunlight, they inhabit the upper, sunlit epipelagic zone, which includes the coastal or neritic zone. Biodiversity diminishes markedly in the deeper zones below the epipelagic zone as dissolved oxygen diminishes, water pressure increases, temperatures become colder, food sources become scarce, and light diminishes and finally disappears.

Pelagic Birds

Pelagic birds, also called oceanic birds, live on the open sea, rather than around waters adjacent to land or around inland waters. Pelagic birds feed on planktonic crustaceans, squid and forage fish. Examples are the Atlantic puffin, macaroni penguins, sooty terns, shearwaters, and procellariiforms such as the albatross, procellariids and petrels.

The term seabird includes birds which live around the sea adjacent to land, as well as pelagic birds.

Pelagic Fish

Pelagic fish live in the water column of coastal, ocean, and lake waters, but not on or near the bottom of the sea or the lake. They can be contrasted with demersal fish, which live on or near the bottom, and reef fish, which are associated with coral reefs.

These fish are often migratory forage fish, which feed on plankton, and the larger fish that follow and feed on the forage fish. Examples of migratory forage fish are herring, anchovies, capelin, and menhaden. Examples of larger pelagic fish which prey on the forage fish are billfish, tuna, and oceanic sharks.

Pelagic Invertebrates

Some examples of pelagic invertebrates include krill, copepods, jellyfish, decapod larvae, hyperiid amphipods, rotifers and cladocerans.

Thorson's rule states that benthic marine invertebrates at low latitudes tend to produce large numbers of eggs developing to widely dispersing pelagic larvae, whereas at high latitudes such organisms tend to produce fewer and larger lecithotrophic (yolk-feeding) eggs and larger offspring.

Pelagic Reptiles

Pelamis platura, the pelagic sea snake, is the only one of the 65 species of marine snakes to spend its entire life in the pelagic zone. It bears live young at sea and is helpless on land. The species sometimes forms aggregations of thousands along slicks in surface waters. The pelagic sea snake is the world's most widely distributed snake species.

Many species of sea turtles spend the first years of their lives in the pelagic zone, moving closer to shore as they reach maturity.

Tide Pool

Tide pools, or rock pools, are rocky pools on the sea shore which are filled with seawater. Many of these pools exist as separate pools only at low tide.

The side of a tide pool in Santa Cruz, California showing sea stars (*Dermasterias*), sea anemones (*Anthopleura*) and sea sponges.

A tide pool in Porto Covo, west coast of Portugal

Many tide pools are habitats of especially adaptable animals that have engaged the attention of naturalists and marine biologists, as well as philosophical essayists: John Steinbeck wrote in *The Log from the Sea of Cortez*, "It is advisable to look from the tide pool to the stars and then back to the tide pool."

Zones from Shallow to Deep

Tide pools in Santa Cruz, California from spray/splash zone to low tide zone

Tidal pools exist in the intertidal zones. These zones are submerged by the sea at high tides and during storms, and may receive spray from wave action. At other times the rocks may undergo other extreme conditions, baking in the sun or exposed to cold winds. Few organisms can survive such harsh conditions. Lichens and barnacles live in this region. In this zone, different barnacle species live at very tightly constrained elevations. Tidal conditions precisely determine the exact height of an assemblage relative to sea level.

The intertidal zone is periodically exposed to sun and wind, which desiccate barnacles, which need to be well adapted to water loss. Their calcite shells are impermeable, and they possess two plates which they slide across their mouth opening when not feeding. These plates also protect against predation.

High Tide Zone

The high tide zone is flooded during each high tide. Organisms must survive wave action, currents, and exposure to the sun. This zone is predominantly inhabited by seaweed and invertebrates, such as sea anemones, starfish, chitons, crabs, green algae, and mussels. Marine algae provide shelter for nudibranches and hermit crabs. The same waves and currents that make life in the high tide zone difficult bring food to filter feeders and other intertidal organisms.

Low tide zone in a tide pool

Low Tide Zone

Also called the Lower Littoral Zone. This area is usually under water - it is only exposed when the tide is unusually low. This sub region is mostly submerged, but it is exposed only during low tide. Often it teems with life and has much more marine vegetation, especially seaweeds. There is also greater biodiversity. Organisms in this zone do not have to be as well adapted to drying out and temperature extremes. Low tide zone organisms include abalone, anemones, brown seaweed, chitons, crabs, green algae, hydroids, isopods, limpets, and mussels. These creatures can grow to larger sizes because there is more available energy and better water coverage: the water is shallow enough to allow more sunlight for photosynthetic activity, and the salinity is at almost normal levels. This area is also relatively protected from large predators because of the wave action and shallow water.

Life in the Tide Pool

Tide pools provide a home for hardy organisms such as starfish, mussels and clams. Inhabitants must be able to deal with a frequently changing environment — fluctuations in water temperature, salinity, and oxygen content. Hazards include waves, strong currents, exposure to midday sun and predators.

Waves can dislodge mussels and draw them out to sea. Gulls pick up and drop sea urchins to break them open. Starfish prey on mussels and are eaten by gulls themselves. Even black bears sometimes feast on intertidal creatures at low tide. Although tide pool organisms must avoid getting washed away into the ocean, drying up in the sun, or getting eaten, they depend on the tide pool's constant changes for food.

Fauna

The sea anemone *Anthopleura elegantissima* reproduces clones of itself through a process called longitudinal fission, in which the animal splits into two parts along its length. The sea anemone *Anthopleura sola* often engages in territorial fights. The white tentacles (acrorhagi), which contain stinging cells, are for fighting. The sea anemones sting each other repeatedly until one moves.

Some species of starfish can regenerate lost arms. Most species must retain an intact central part of the body to be able to regenerate, but a few can regrow from a single ray. The regeneration of these stars is possible because the vital organs are in the arms.

Flora

Sea palms look similar to palm trees. They live in the middle to upper intertidal zones in areas with greater wave action. High wave action may increase nutrient availability and moves the blades of the thallus, allowing more sunlight to reach the organism so that it can photosynthesize. In addition, the constant wave action removes competitors, such as the mussel species *Mytilus californianus*.

Recent studies have shown that *Postelsia* grows in greater numbers when such competition exists — a control group with no competition produced fewer offspring than an experimental group with mussels; from this it is thought that the mussels provide protection for the developing gametophytes. Alternatively, the mussels may prevent the growth of competing algae such as *Corallina* or *Halosaccion*, allowing *Postelsia* to grow freely after wave action removes the mussels.

A large sea anemone *Anthopleura sola* consuming a "by-the-wind-sailor" *Velella velella* a blue hydrozoan

Postelsia palmaeformis at low tide in a tide pool

Sea star, *Pisaster ochraceus* consuming a mussel in tide pools

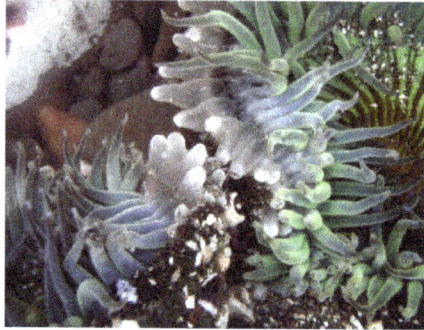

Sea anemones, *Anthopleura sola* engaged in a battle for territory

References

- Silva PC, Basson PW and Moe RL (1996) Catalogue of the Benthic Marine Algae of the Indian Ocean page 2, University of California Press. ISBN 9780520915817.

- McLusky, D. S.; Elliott, M. (2004). The Estuarine Ecosystem: Ecology, Threats and Management. New York: Oxford University Press. ISBN 0-19-852508-7.

- Gould, Stephen Jay (1990) Wonderful Life: The Burgess Shale and the Nature of History W. W. Norton. ISBN 9780393307009.

- Ponder, W.F.; Lindberg, D.R., eds. (2008). Phylogeny and Evolution of the Mollusca. Berkeley: University of California Press. p. 481. ISBN 978-0-520-25092-5.

- Karleskint G, Richard Turner R and , James Small J (2012) Introduction to Marine Biology Cengage Learning, edition 4, page 445. ISBN 9781133364467.

- Platt HM (1994). "foreword". In Lorenzen S, Lorenzen SA. The phylogenetic systematics of freeliving nematodes. London: The Ray Society. ISBN 0-903874-22-9.

- Wilbur, Karl M.; Clarke, M.R.; Trueman, E.R., eds. (1985), The Mollusca, 12. Paleontology and neontology of Cephalopods, New York: Academic Press, ISBN 0-12-728702-7

- Ruppert, Edward E.; Fox, Richard, S.; Barnes, Robert D. (2004). Invertebrate Zoology, 7th edition. Cengage Learning. ISBN 81-315-0104-3.

- Hayward, PJ (1996). Handbook of the Marine Fauna of North-West Europe. Oxford University Press. ISBN 0-19-854055-8.

- Healy, J.M. (2001). "The Mollusca". In Anderson, D.T. Invertebrate Zoology (2 ed.). Oxford University Press. pp. 120–171. ISBN 0-19-551368-1.

- Walker P and Wood E (2005) The Open Ocean (volume in a series called Life in the sea), Infobase Publishing, ISBN 978-0-8160-5705-4.

- Lal BV and Fortune K (2000) The Pacific Islands: An encyclopedia Page 8. University of Hawaii Press. ISBN 978-0-8248-2265-1.

- Spalding, Mark, Corinna Ravilious, and Edmund Green (2001). World Atlas of Coral Reefs. Berkeley, CA: University of California Press and UNEP/WCMC ISBN 0520232550.

- Moyle, Peter B.; Joseph J. Cech (2004). Fishes : an introduction to ichthyology (Fifth ed.). Upper Saddle River, N.J.: Pearson/Prentice Hall. p. 556. ISBN 978-0-13-100847-2.

- Marshall, Paul; Schuttenberg, Heidi (2006). A Reef Manager's Guide to Coral Bleaching. Townsville, Australia: Great Barrier Reef Marine Park Authority. ISBN 1-876945-40-0.

- Osborne, Patrick L. (2000). Tropical Ecosystem and Ecological Concepts. Cambridge: Cambridge University Press. p. 464. ISBN 0-521-64523-9.

- Lutz, Peter L.; Musick, John A. (1996). The biology of sea turtles. Boca Raton, Fla: CRC Press. ISBN 0849384222.

- Jennings S, Kaiser MJ and Reynolds JD (2001) Marine fisheries ecology, Wiley-Blackwell, pp. 291–293. ISBN 978-0-632-05098-7.

Understanding Marine Toxicology

To have a precise understanding of marine toxicology, it is very important to understand aquatic toxicology and cyanotoxin. Industrial waste and manufactured chemicals have an immense effect on marine biology. The study of these effects is known as aquatic toxicology. This section has been carefully written to provide an easy understanding of marine toxicology.

Aquatic Toxicology

Aquatic toxicology is the study of the effects of manufactured chemicals and other anthropogenic and natural materials and activities on aquatic organisms at various levels of organization, from subcellular through individual organisms to communities and ecosystems. Aquatic toxicology is a multidisciplinary field which integrates toxicology, aquatic ecology and aquatic chemistry.

This field of study includes freshwater, marine water and sediment environments. Common tests include standardized acute and chronic toxicity tests lasting 24–96 hours (acute test) to 7 days or more (chronic tests). These tests measure endpoints such as survival, growth, reproduction, that are measured at each concentration in a gradient, along with a control test. Typically using selected organisms with ecologically relevant sensitivity to toxicants and a well-established literature background. These organisms can be easily acquired or cultured in lab and are easy to handle.

History

While basic research in toxicology began in multiple countries in the 1800s, it was not until around the 1930s that the use of acute toxicity testing, especially on fish, was established. Over the next two decades, the effects of chemicals and wastes on non-human species became more of a public issue and the era of the *pickle-jar bioassays* began as efforts increased to standardize toxicity testing techniques. In the United States of America, the passage of the Federal Water Pollution Control Act of 1947 marked the first comprehensive legislation for the control of water pollution and was followed by the Federal Water Pollution Control Act in 1956. In 1962, public and governmental interests were renewed, in large part due to the publication of Rachel Carson's *Silent Spring*, and three years later the Water Quality Act was passed which directed states to develop water quality standards. Public awareness, as well as scientific and governmental concern, continued to grow throughout the 1970s and by the end of the decade research had expanded to include hazard evaluation and risk analysis. In the subsequent decades, aquatic toxicology has continued to expand and internationalize so that there is now a strong application of toxicity testing for environmental protection.

Aquatic Toxicity Tests

Aquatic toxicology tests (assays): toxicity tests are used to provide qualitative and quantitative data on adverse (deleterious) effects on aquatic organisms from a toxicant. Toxicity tests can be used to assess the potential for damage to an aquatic environment and provide a database that can be used to assess the risk associated within a situation for a specific toxicant. Aquatic toxicology tests can be performed in the field or in the laboratory. Field experiments generally refer to multiple species exposure and laboratory experiments generally refer to single species exposure. A dose response relationship is most commonly used with a sigmoidal curve to quantify the toxic effects at a selected end-point or criteria for effect (i.e. death or other adverse effect to the organism). Concentration is on the x-axis and percent inhibition or response is on the y-axis.

The criteria for effects, or endpoints tested for, can include lethal and sublethal effects.

There are different types of toxicity tests that can be performed on various test species. Different species differ in their susceptibility to chemicals, most likely due to differences in accessibility, metabolic rate, excretion rate, genetic factors, dieteary factors, age, sex, health and stress level of the organism. Common standard test species are the fathead minnow (Pimephales promelas), daphnids (*Daphnia magna, D. pulex, D. pulicaria, Ceriodaphnia dubia*), midge (Chironomus tentans, C. ruparius), rainbow trout (Oncorhynchus mykiss), sheepshead minnow (Cyprinodon variegatu), mysids (Mysidopsis), oyster (Crassotreas), scud (Hyalalla Azteca), grass shrimp (Palaemonetes pugio), mussels (Mytilus). As defined by ASTM, these species are routinely selected on the basis of availability, commercial, recreational, and ecological importance, past successful use, and regulatory use.

A variety of acceptable standardized test methods have been published. Some of the more widely accepted agencies to publish methods are: the American Public Health Association, U.S. Environmental Protection Agency, American Society for Testing and Materials, International Organization for Standardization, Environment Canada, and Organization for Economic Cooperation and Development. Standardized tests offer the ability to compare results between laboratories.

There are many kinds of toxicity tests widely accepted in the scientific literature and regulatory agencies. The type of test used depends on many factors: Specific regulatory agency conducting the test, resources available, physical and chemical characteristics of the environment, type of toxicant, test species available, laboratory vs. field testing, end-point selection, and time and resources available to conduct the assays are some of the most common influencing factors on test design.

Exposure Systems

Exposure systems are four general techniques the controls and test organisms are exposed to the dealing with treated and diluted water or the test solutions.

Static- a static test exposes the organism in still water. The toxicant is added to the water in order to obtain the correct concentrations to be tested. The control and test organisms are placed in the test solutions and the water is not changed for the entirety of the test.

Recirculation- a recirculation test exposes the organism to the toxicant in a similar manner as the static test, except that the test solutions are pumped through an apparatus (i.e. filter) to maintain water quality, but not reduce the concentration of the toxicant in the water. The water is circulated through the test chamber continuously, similar to an aerated fish tank. This type of test is expensive and it is unclear whether or not the filter or aerator has an effect on the toxicant.

Renewal- a renewal test also exposes the organism to the toxicant in a similar manner as the static test because it is in still water. However, in a renewal test the test solution is renewed periodically (constant intervals) by transferring the organism to a fresh test chamber with the same concentration of toxicant.

Flow-through- a flow through test exposes the organism to the toxicant with a flow into the test chambers and then out of the test chambers. The once-through flow can either be intermittent or continuous. A stock solution of the correct concentrations of contaminant must be previously prepared. Metering pumps or diluters will control the flow and the volume of the test solution, and the proper proportions of water and contaminant will be mixed.

Types of Tests

Acute tests are short-term exposure tests (hours or days) and generally use lethality as an endpoint. In acute exposures, organisms come into contact with higher doses of the toxicant in a single event or in multiple events over a short period of time and usually produce immediate effects, depending on absorption time of the toxicant. These tests are generally conducted on organisms during a specific time period of the organism's life cycle, and are considered partial life cycle tests. Acute tests are not valid if mortality in the control sample is greater than 10%. Results are reported in EC50, or concentration that will affect fifty percent of the sample size.

Chronic tests are long-term tests (weeks, months years), relative to the test organism's life span (>10% of life span), and generally use sub-lethal endpoints. In chronic exposures, organisms come into contact with low, continuous doses of a toxicant. Chronic exposures may induce effects to acute exposure, but can also result in effects that develop slowly. Chronic tests are generally considered full life cycle tests and cover an entire generation time or reproductive life cycle ("egg to egg"). Chronic tests are not considered valid if mortality in the control sample is greater than 20%. These results are generally reported in NOECs (No observed effects level) and LOECs (Lowest observed effects level).

Early life stage tests are considered as subchronic exposures that are less than a complete reproductive life cycle and include exposure during early, sensitive life stages of an organism. These exposures are also called critical life stage, embryo-larval, or egg-fry tests. Early life stage tests are not considered valid if mortality in the control sample is greater than 30%.

Short-term sublethal tests are used to evaluate the toxicity of effluents to aquatic organisms. These methods are developed by the EPA, and only focus on the most sensitive life stages. Endpoints for these test include changes in growth, reproduction and survival. NOECs, LOECs and EC50s are reported in these tests.

Bioaccumulation tests are toxicity tests that can be used for hydrophobic chemicals that may accumulated in the fatty tissue of aquatic organisms. Toxicants with low solubilities in water generally can be stored in the fatty tissue due to the high lipid content in this tissue. The storage of these toxicants within the organism may lead to cumulative toxicity. Bioaccumulation tests use bioconcentration factors (BCF) to predict concentrations of hydrophobic contaminants in organisms. The BCF is the ratio of the average concentration of test chemical accumulated in the tissue of the test organism (under steady state conditions) to the average measured concentration in the water.

Freshwater tests and saltwater tests have different standard methods, especially as set by the regulatory agencies. However, these tests generally include a control (negative and/or positive), a geometric dilution series or other appropriate logarithmic dilution series, test chambers and equal numbers of replicates, and a test organism. Exact exposure time and test duration will depend on type of test (acute vs. chronic) and organism type. Temperature, water quality parameters and light will depend on regulator requirements and organism type.

Effluent toxicity tests are tests conducted under the Clean Water Act, National Pollutant Discharge Elimination System (NPDES) permit program and are used by dischargers of contaminated effluent to monitor the quality of effluent into receiving waters. Acute Effluent Toxicity Tests are used to monitor the quality of industrial effluent monthly using acute toxicity tests. Effluent is used to perform static-acute multi concentration toxicity tests with *Ceriodaphnia dubia* and *Pimephales promelas*. The test organisms are exposed for 48 hours under static conditions with five concentrations of the effluent. Short-term Chronic Effluent Toxicity Tests are used to monitor the quality of municipal wastewater treatment plants effluent quarterly using short-term chronic toxicity tests. The goal of this test is to ensure that the wastewater is not chronically toxic. The major deviation in the short-term chronic effluent toxicity tests and the acute effluent toxicity tests is that the short-term chronic test lasts for seven days and the acute test lasts for 48 hours.

Sediment Tests

At some point most chemicals originating from both anthropogenic and natural sources accumulate in sediment. For this reason, sediment toxicity can play a major role in the adverse biological effects seen in aquatic organisms, especially those inhabiting benthic habitats. A recommended approach for sediment testing is to apply the Sediment Quality Triad (SQT) which involves simultaneously examining sediment chemistry, toxicity, and field alterations so that more complete information can be gathered. Collection, handling, and storage of sediment can have an effect on bioavailability and for this reason standard methods have been developed to suit this purpose.

Toxicological Effects

Toxicity can be broken down into two broad categories of direct and indirect toxicity. Direct toxicity results from a toxicant acting at the site of action in or on the organism. Indirect toxicity occurs with a change in the physical, chemical, or biological environment.

Lethality is most common effect used in toxicology and used as an endpoint for acute toxicity tests. While conducting chronic toxicity tests sublethal effects are endpoints that are looked at. These endpoints include behavioral, physiological, biochemical, histological changes.

There are a number of effects that occur when an organism is simultaneously exposed to two or more toxicants. These effects include additive effects, synergistic effects, potentiation effects, and antagonistic effects. An additive effect occurs when combined effect is equal to a combination or sum of the individual effects. A synergistic effect occurs when the combination of effects is much greater than the two individual effects added together. Potentiation is an effect that occurs when an individual chemical has no effect is added to a toxicant and the combination has a greater effect than just the toxicant alone. Finally, an antagonistic effect occurs when a combination of chemicals has less of an effect than the sum of their individual effects.

Important Aquatic Toxicology Resources

- American Society for Testing and Materials (ASTM International) – A consensus organization, representing 135 countries, that develops and delivers international voluntary standard methods for aquatic toxicity testing.

- Standard Methods for the Examination of Water and Wastewater – A compilation of techniques for the examination of water, jointly published by the American Public Health Association (APHA), the American Water Works Association (AWWA), and the Water Pollution Control Federation (WPCF).

- Ecotox – A database maintained by the U.S. Environmental Protection Agency (EPA) that offers single chemical toxicity information for both aquatic and terrestrial purposes.

- Society of Environmental Toxicology and Chemistry (SETAC) – A nonprofit, worldwide society working to promote scientific research to further our understanding of environmental stressors, environmental education, and the use of science in environmental policy.

- United States Environmental Protection Agency (EPA) – A federal agency working to protect human and environmental health. Among many other functions, the U.S. EPA produces guidance manuals outlining aquatic toxicity test procedures.

- Organisation for Economic Co-operation and Development (OECD) – A forum for governments to work together to promote policies for the betterment of people's social and economic well-being around the world. One way in which they accomplish this is through the development of aquatic toxicity test guidelines.

- Environment Canada (EC) – A diverse organization working to protect Canada's water resources and the natural environment through the coordination of environmental policies and programs with the federal government.

Terminology

- Median Lethal Concentration (LC50) – The chemical concentration that is expected to kill 50% of a group of organisms.

- Median Effective Concentration (EC50) – The chemical concentration that is expected to have one or more specified effects in 50% of a group of organisms.

- Critical Body Residue (CBR) – An approach that routinely examines whole-body chemical concentrations of an exposed organism that is associated with an adverse biological response.

- Baseline toxicity – Refers to narcosis which is a depression in biological activity due to toxicants being present in the organism.

- Biomagnification – The process by which the concentration of a chemical in the tissues of an organism increases as it passes through several levels in the food web.

- Lowest Observed Effect Concentration (LOEC) – The lowest test concentration that has a statistically significant effect over a specified exposure time.

- No Observed Effect Concentration (NOEC) – The highest test concentration for which no effect is observed relative to a control over a specified exposure time.

- Maximum Acceptable Toxicant Concentration (MATC) – An estimated value that represents the highest "no-effect" concentration of a specific substance within the range including the NOEC and LOEC.

- Application Factor (AF) – An empirically derived "safe" concentration of a chemical.

- Biomonitoring – The consistent use of living organisms to analyze environmental changes over time.

- Effluent – Liquid, industrial discharge that usually contain varying chemical toxicants.

- Quantitative Structure-Activity Relationship (QSAR) – A method of modeling the relationship between biological activity and the structure of organic chemicals.

- Mode of Action – A set of common behavioral or physiological signs that represent a type of adverse response.

- Mechanism of Action – The detailed events that take place at the molecular level during an adverse biological response.

- KOW – The octanol-water partition coefficient which represents the ratio of the concentration of octanol to the concentration of chemical in the water.

- Bioconcentration Factor (BCF) – The ratio of the average chemical concentration in the tissues of the organism under steady-state conditions to the average chemical concentration measured in the water to which the organisms are exposed.

All terms were derived from Rand.

Significance to Regulatory World

In the United States aquatic toxicology plays an important role in the NPDES wastewater permit program. In addition to analytical testing for known pollutants, aquatic, whole effluent toxicity tests have been standardized and are performed routinely as a tool for evaluating the potential harmful effects of effluents discharged into surface waters.

For the Clean Water Act under United States Environmental Protection Agency there are water quality criteria and water quality standards derived from aquatic toxicity tests.

Sediment Quality Guidelines

While sediment quality guidelines are not meant for regulation, they provide a way to rank and compare sediment quality developed by National Oceanic and Atmospheric Administration(NOAA). These sediment quality guidelines are summarized in NOAA's Screening Quick Reference Tables (SQuiRT) for many different chemicals.

Cyanotoxin

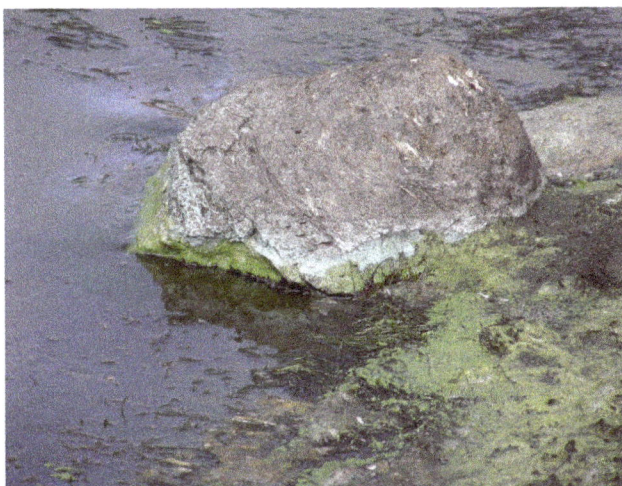

Green scum produced by and containing cyanobacteria, washed up on a rock
in California during an algal bloom

Cyanotoxins are toxins produced by bacteria called cyanobacteria (also known as blue-green algae). Cyanobacteria are found almost everywhere, but particularly in lakes and in the ocean where, under certain conditions, they reproduce exponentially to form blooms. Blooming cyanobacteria can produce cyanotoxins in such concentrations that they poison and even kill animals and humans. Cyanotoxins can also accumulate in other animals such as fish and shellfish, and cause poisonings such as shellfish poisoning.

Among cyanotoxins are some of the most powerful natural poisons known, including poisons which can cause rapid death by respiratory failure. The toxins include potent neurotoxins, hepatotoxins, cytotoxins, and endotoxins. Despite the similarity in name, they are not cyanides. Recreational exposure to cyanobacteria can result in gastro-intestinal and hay fever symptoms or pruritic skin rashes. Exposure to the cyanobacteria neurotoxin BMAA may be an environmental cause of neuro-degenerative diseases such as ALS, Parkinson's Disease and Alzheimer's Disease. There is also an interest in the military potential of biological neurotoxins such as cyanotoxins, which "have gained increasing significance as potential candidates for weaponization."

The first published report that blue-green algae or cyanobacteria could have lethal effects appeared in *Nature* in 1878. George Francis described the algal bloom he observed in the estuary of the Murray River in Australia, as "a thick scum like green oil paint, some two to six inches thick." Wildlife which drank the water died rapidly and terribly. Most reported incidents of poisoning by microalgal toxins have occurred in freshwater environments, and they are becoming more common and widespread. For example, thousands of ducks and geese died drinking contaminated water in the midwestern United States. In 2010, for the first time, marine mammals were reported to have died from ingesting cyanotoxins.

Cyanobacteria

Cyanotoxins are produced by cyanobacteria, a phylum of bacteria that obtain their energy through photosynthesis. The prefix *cyan* comes from the Greek κύανος meaning "a dark blue substance", and usually indicates any of a number of colours in the blue/green range of the spectrum. Cyanobacteria are commonly referred to as *blue-green algae*. Traditionally they were thought of as a form of algae, and were introduced as such in older textbooks. However modern sources tend to regard this as outdated; they are now considered to be more closely related to bacteria, and the term for true *algae* is restricted to eukaryotic organisms. Like true algae, cyanobacteria are photosynthetic and contain photosynthetic pigments, which is why they are usually green or blue.

Cyanobacteria are found almost everywhere; in oceans, lakes and rivers as well as on land. They flourish in Arctic and Antarctic lakes, hotsprings and wastewater treatment plants. They even inhabit the fur of polar bears, to which they impart a greenish tinge. Cyanobacteria produce potent toxins, but they also produce helpful bioactive compounds, including substances with antitumour, antiviral, anticancer, antibiotic and antifungal activity, UV protectants and specific inhibitors of enzymes.

Harmful Algal Blooms

Dense bloom of cyanobacteria on the Potomac River estuary. These blooms can be toxic.

Cyanotoxins are often implicated in what are commonly called *red tides* or *harmful algal blooms*. Lakes and oceans contain many single-celled organisms called phytoplankton. Under certain conditions, particularly when nutrient concentrations are high, these organisms reproduce exponentially. The resulting dense swarm of phytoplankton is called an algal bloom; these can cover hundreds of square kilometres and can be easily seen in satellite images. Individual phytoplankton rarely live more than a few days, but blooms can last weeks.

Generally these blooms are harmless, but if not they are called harmful algal blooms, or HABs. HABs can contain toxins or pathogens which result in fish kill and can also be fatal to humans. In marine environments, HABs are mostly caused by dinoflagellates, though species of other algae taxa can also cause HABs (diatoms, flagellates, haptophytes and raphidophytes). Marine dinoflagellate species are often toxic, but freshwater species are not known to be toxic. Neither are diatoms known to be toxic, at least to humans.

In freshwater ecosystems, algal blooms are most commonly caused by eutrophication. The blooms can look like foam, scum or mats or like paint floating on the surface of the water, but they are not always visible. Nor are the blooms always green; they can be blue, and some cyanobacteria species are coloured brownish-red. The water can smell bad when the cyanobacteria in the bloom die.

Strong cyanobacterial blooms reduce visibility to one or two centimetres. Species which do not need to see to migrate in the water column (such as the cyanobacteria themselves) survive, but species which need to see to find food and partners are compromised. During the day blooming cyanobacteria saturate the water with oxygen. At night respiring aquatic organisms can deplete the oxygen to the point where sensitive species, such as certain fish, die. This is more likely to happen near the sea floor or a thermocline. Water acidity also cycles daily during a bloom, with the pH reaching 9 or more during the day and dropping to low values at night, further stressing the ecosystem. In addition, many cyanobacteria species produce potent cyanotoxins which concentrate during a bloom to the point where they become lethal to nearby aquatic organisms and any other animals in direct contact with the bloom, including birds, livestock, domestic animals and sometimes humans.

In 1991 a harmful cyanobacterial bloom affected 1000 km of the Darling-Barwon River in Australia at an economic cost of $10M AUD.

Chemical Structure

Cyanotoxins usually target the nervous system (neurotoxins), the liver (hepatotoxins) or the skin (dermatoxins). The chemical structure of cyanotoxins falls into three broad groups: cyclic peptides, alkaloids and lipopolysaccharides (endotoxins).

Chemical structure of cyanotoxins

Structure	Cyanotoxin	Primary target organ in mammals	Cyanobacteria genera
Cyclic peptides	Microcystins	Liver	*Microcystis, Anabaena, Planktothrix* (Oscillatoria), *Nostoc, Hapalosiphon, Anabaenopsis*
	Nodularins	Liver	*Nodularia*
Alkaloids	Anatoxin-a	Nerve synapse	*Anabaena, Planktothrix* (Oscillatoria), *Aphanizomenon*
	Anatoxin-a(S)	Nerve synapse	*Anabaena*
	Cylindrospermopsins	Liver	*Cylindrospermopsis, Aphanizomenon, Umezakia*
	Lyngbyatoxin-a	Skin, gastro-intestinal tract	*Lyngbya*
	Saxitoxin	Nerve synapse	*Anabaena, Aphanizomenon, Lyngbya, Cylindrospermopsis*
	Lipopolysaccharides	Potential irritant; affects any exposed tissue	All
Polyketides	Aplysiatoxins	Skin	*Lyngbya, Schizothrix, Planktothrix* (Oscillatoria)
Amino Acid	BMAA	Nervous System	All

Most cyanotoxins have a number of variants (analogues). As of 1999, altogether over 84 cyanotoxins were known and only a small number have been well studied.

Cyclic Peptides

A peptide is a short polymer of amino acids linked by peptide bonds. They have the same chemical structure as proteins, except they are shorter. In a cyclic peptide the links link back to the start to form a stable circular chain. In mammals this stability makes them resistant to the process of digestion and they can bioaccumulate in the liver. Of all the cyanotoxins, the cyclic peptides are of most concern to human health. The microcystins and nodularins poison the liver, and exposure to high doses can cause death. Exposure to low doses in drinking water over a long period of time may promote liver and other tumours.

Microcystins

As with other cyanotoxins, microcystins were named after the first organism discovered to produce them, *Microcystis aeruginosa*. However it was later found other cyanobacterial genera also produced them. There are about 60 known variants of microcystin, and several of these can be produced during a bloom. The most reported variant is microcystin-LR, possibly because the earliest commercially available chemical standard analysis was for microcystin-*LR*.

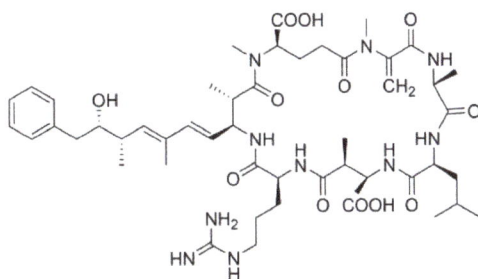

Microcystin LR

Blooms containing microcystin are a problem worldwide in freshwater ecosystems. Microcystins are cyclic peptides and can be very toxic for plants and animals including humans. They bioaccumulate in the liver of fish, in the hepatopancreas of mussels, and in zooplankton. They are hepatotoxic and can cause serious damage to the liver in humans. In this way they are similar to the nodularins (below), and together the microcystins and nodularins account for most of the toxic cyanobacterial blooms in fresh and brackish waters. In 2010, a number of sea otters were poisoned by microcystin. Marine bivalves were the likely source of hepatotoxic shellfish poisoning. This was the first confirmed example of a marine mammal dying from ingesting a cyanotoxin.

Nodularins

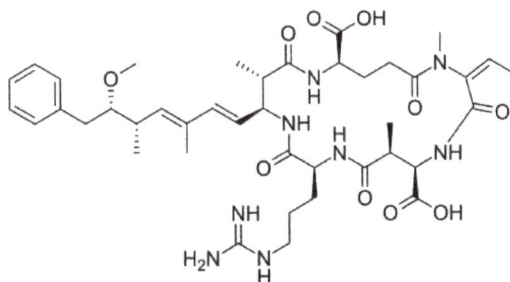

Nodularin-R

The first nodularin variant to be identified was nodularin-R, produced by the cyanobacterium *Nodularia spumigena*. This cyanobacterium blooms in water bodies throughout the world. In the Baltic Sea, marine blooms of *Nodularia spumigena* are among some of the largest cyanobacterial mass events in the world.

Globally, the most common toxins present in cyanobacterial blooms in fresh and brackish waters are the cyclic peptide toxins of the nodularin family. Like the microcystin family (above), nodularins are potent hepatotoxins and can cause serious damage to the liver. They present health risks for wild and domestic animals as well as humans, and in many areas pose major challenges for the provision of safe drinking water.

Alkaloids

Alkaloids are a group of naturally occurring chemical compounds which mostly contain basic nitrogen atoms. They are produced by a large variety of organisms, including cyanobacteria, and are part of the group of natural products, also called secondary metabolites. Alkaloids act on diverse metabolic systems in humans and other animals, often with psychotropic or toxic effects. Almost uniformly, they are bitter tasting.

Anatoxin-*a*

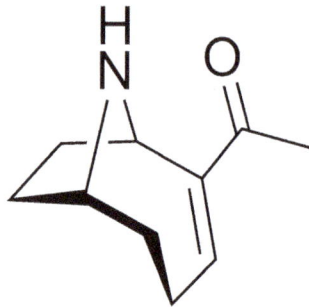

Anatoxin-*a*

Investigations into anatoxin-*a*, also known as "Very Fast Death Factor", began in 1961 following the deaths of cows that drank from a lake containing an algal bloom in Saskatchewan, Canada. The toxin is produced by at least four different genera of cyanobacteria and has been reported in North America, Europe, Africa, Asia, and New Zealand.

Toxic effects from anatoxin-*a* progress very rapidly because it acts directly on the nerve cells (neurons) as a neurotoxin. The progressive symptoms of anatoxin-*a* exposure are loss of coordination, twitching, convulsions and rapid death by respiratory paralysis. The nerve tissues which communicate with muscles contain a receptor called the nicotinic acetylcholine receptor. Stimulation of these receptors causes a muscular contraction. The anatoxin-*a* molecule is shaped so it fits this receptor, and in this way it mimics the natural neurotransmitter normally used by the receptor, acetylcholine. Once it has triggered a contraction, anatoxin-*a* does not allow the neurons to return to their resting state, because it is not degraded by cholinesterase which normally performs this function. As a result, the muscle cells contract permanently, the communication between the brain and the muscles is disrupted and breathing stops.

The toxin was called the Very Fast Death Factor because it induced tremors, paralysis and death within a few minutes when injected into the body cavity of mice. In 1977, the structure of VFDF was determined as a secondary, bicyclic amine alkaloid, and it was renamed anatoxin-a. Structurally, it is similar to cocaine. There is continued interest in anatoxin-a because of the dangers it presents to recreational and drinking waters, and because it is a particularly useful molecule for investigating acetylcholine receptors in the nervous system. The deadliness of the toxin means that it has a high military potential as a toxin weapon.

Cylindrospermopsins

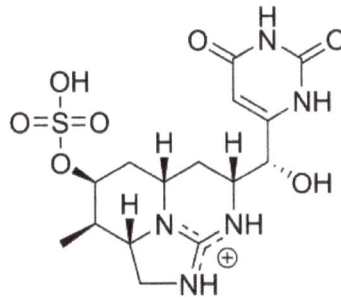

Cylindrospermopsin

Cylindrospermopsin (abbreviated to CYN or CYL) was first discovered after an outbreak of a mystery disease on Palm Island in Australia. The outbreak was traced back to a bloom of *Cylindrospermopsis raciborskii* in the local drinking water supply, and the toxin was subsequently identified. Analysis of the toxin led to a proposed chemical structure in 1992, which was revised after synthesis was achieved in 2000. Several variants of cylindrospermopsin, both toxic and non-toxic, have been isolated or synthesised.

Cylindrospermopsin is toxic to liver and kidney tissue and is thought to inhibit protein synthesis and to covalently modify DNA and/or RNA. There is concern about the way cylindrospermopsin bioaccumulates in freshwater organisms. Toxic blooms of genera which produce cylindrospermopsin are most commonly found in tropical, subtropical and arid zone water bodies, and have recently been found in Australia, Europe, Israel, Japan and the USA.

Saxitoxins

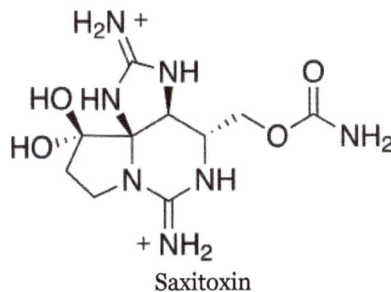

Saxitoxin

Saxitoxin (STX) is one of the most potent natural neurotoxins known. The term saxitoxin originates from the species name of the butter clam (*Saxidomus giganteus*) whereby it was first recognized. Saxitoxin is produced by the cyanobacteria *Anabaena* spp., some *Aphanizomenon* spp., *Cylindrospermopsis* sp., *Lyngbya* sp. and *Planktothrix* sp.). Puffer fish and some marine dinoflagellates also produce saxitoxin. Saxitoxins bioaccumulate in shellfish and certain finfish. Ingestion of saxitoxin, usually through shellfish contaminated by toxic algal blooms, can result in paralytic shellfish poisoning.

Saxitoxin has been used in molecular biology to establish the function of the sodium channel. It acts on the voltage-gated sodium channels of nerve cells, preventing normal cellular function and leading to paralysis. The blocking of neuronal sodium channels which occurs in paralytic shellfish poisoning produces a flaccid paralysis that leaves its victim calm and conscious through the progression of symptoms. Death often occurs from respiratory failure. Saxitoxin was originally isolated and described by the United States military, who assigned it the chemical weapon designation "TZ". Saxitoxin is listed in schedule 1 of the Chemical Weapons Convention. According to the book *Spycraft*, U-2 spyplane pilots were provided with needles containing saxitoxin to be used for suicide in the event escape was impossible.

Lipopolysaccharides

Lipopolysaccharides are present in all cyanobacteria. Though not as potent as other cyanotoxins, some researchers have claimed that all lipopolysaccharides in cyanobacteria can irritate the skin, while other researchers doubt the toxic effects are that generalized.

Amino Acids

BMAA

The non-proteinogenic amino acid beta-Methylamino-L-alanine (BMAA) is ubiquitously produced by cyanobacteria in marine, freshwater, brackish, and terrestrial environments. The exact mechanisms of BMAA toxicity on neuron cells is being investigated. Research suggests both acute and chronic mechanisms of toxicity. BMAA is being investigated as a potential environmental risk factor for neurodegenerative diseases, including ALS, Parkinson's disease and Alzheimer's disease.

References

- Rand, Gary M.; Petrocelli, Sam R. (1985). Fundamentals of aquatic toxicology: Methods and applications. Washington: Hemisphere Publishing. ISBN 0-89116-382-4.

- Chorus I and Bartram J (1999) Toxic cyanobacteria in water: A guide to their public health consequences, monitoring and management World Health Organisation. E & FN Spon, ISBN 0-419-23930-8.

- Pelaez M et al. (2010) "Sources and Occurrence of Cyanotoxins Worldwide". In Xenobiotics in the Urban Water Cycle, Environmental Pollution, 16(I): 101-127, doi:10.1007/978-90-481-3509-7_6

Marine Life: An Overview

Marine life signifies the life of all the organisms and plants that live in the ocean. The area inhabited by marine species is known as marine habitat. Marine habitats are divided into coastal habitats and open ocean habitats. The topics covered in this chapter are very important in developing a complete understanding of the subject.

Marine Life

Marine life refers to the plants, animals and other organisms that live in the ocean. At a fundamental level, marine life helps determine the very nature of our planet. Marine organisms produce much of the oxygen we breathe and probably help regulate the earth's climate. Shorelines are in part shaped and protected by marine life, and some marine organisms even help create new land.

Marine life ranges in size from the microscopic, including plankton and phytoplankton which can be as small as 0.02 micrometres and are both important as key primary producers of the sea, to huge cetaceans (whales, dolphins and porpoises) which in the case of the blue whale reach up to 33 metres (109 feet) in length.

Marine life is an object of study both in marine biology and in biological oceanography. In biology many phyla, families and genera have some species that live in the sea and others that live on land. Marine biology classifies species based on the environment rather than on taxonomy. For this reason marine biology encompasses not only organisms that live only in a marine environment, but also other organisms whose lives revolve around the sea. Biological oceanography is the study of how organisms affect and are affected by the physics, chemistry, and geology of the oceanographic system. Biological oceanography mostly focuses on the microorganisms within the ocean; looking at how they are affected by their environment and how that affects larger marine creatures and their ecosystem. Biological oceanography is similar to marine biology, but is different because of the perspective used to study the ocean. Biological oceanography takes a bottom up approach (in terms of the food web), while marine biology studies the ocean from a top down perspective. Biological oceanography mainly focuses on the ecosystem of the ocean with an emphasis on plankton: their diversity (morphology, nutritional sources, motility, and metabolism); their productivity and how that plays a role in the global carbon cycle; and their distribution (predation and life cycle). Biological oceanography also investigates the role of microbes in food webs, and how humans impact the ecosystems in the oceans.

Evolution

The Earth is about 4.54 billion years old. The earliest undisputed evidence of life on Earth dates from at least 3.5 billion years ago, during the Eoarchean Era after a geological crust started to solidify following the earlier molten Hadean Eon. Microbial mat fossils have been found in 3.48 billion-year-old sandstone in Western Australia. Other early physical evidence of a biogenic substance is graphite in 3.7 billion-year-old metasedimentary rocks discovered in Western Greenland as well as "remains of biotic life" found in 4.1 billion-year-old rocks in Western Australia. According to one of the researchers, "If life arose relatively quickly on Earth … then it could be common in the universe."

All organisms on Earth are descended from a common ancestor or ancestral gene pool. Highly energetic chemistry is thought to have produced a self-replicating molecule around 4 billion years ago, and half a billion years later the last common ancestor of all life existed. The current scientific consensus is that the complex biochemistry that makes up life came from simpler chemical reactions. The beginning of life may have included self-replicating molecules such as RNA and the assembly of simple cells.

Current species are a stage in the process of evolution, with their diversity the product of a long series of speciation and extinction events. The common descent of organisms was first deduced from four simple facts about organisms: First, they have geographic distributions that cannot be explained by local adaptation. Second, the diversity of life is not a set of completely unique organisms, but organisms that share morphological similarities. Third, vestigial traits with no clear purpose resemble functional ancestral traits and finally, that organisms can be classified using these similarities into a hierarchy of nested groups—similar to a family tree. However, modern research has suggested that, due to horizontal gene transfer, this "tree of life" may be more complicated than a simple branching tree since some genes have spread independently between distantly related species.

Past species have also left records of their evolutionary history. Fossils, along with the comparative anatomy of present-day organisms, constitute the morphological, or anatomical, record. By comparing the anatomies of both modern and extinct species, paleontologists can infer the lineages of those species. However, this approach is most successful for organisms that had hard body parts, such as shells, bones or teeth. Further, as prokaryotes such as bacteria and archaea share a limited set of common morphologies, their fossils do not provide information on their ancestry.

More recently, evidence for common descent has come from the study of biochemical similarities between organisms. For example, all living cells use the same basic set of nucleotides and amino acids. The development of molecular genetics has revealed the record of evolution left in organisms' genomes: dating when species diverged through the molecular clock produced by mutations. For example, these DNA sequence comparisons have revealed that humans and chimpanzees share 98% of their genomes and analysing the few areas where they differ helps shed light on when the common ancestor of these species existed.

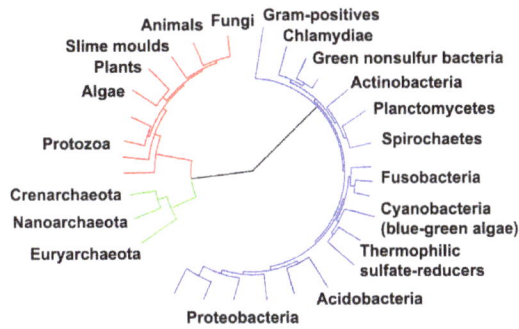

Evolutionary tree showing the divergence of modern species from their common ancestor in the centre. The three domains are coloured, with bacteria blue, archaea green and eukaryotes red.

Prokaryotes inhabited the Earth from approximately 3–4 billion years ago. No obvious changes in morphology or cellular organisation occurred in these organisms over the next few billion years. The eukaryotic cells emerged between 1.6–2.7 billion years ago. The next major change in cell structure came when bacteria were engulfed by eukaryotic cells, in a cooperative association called endosymbiosis. The engulfed bacteria and the host cell then underwent coevolution, with the bacteria evolving into either mitochondria or hydrogenosomes. Another engulfment of cyanobacterial-like organisms led to the formation of chloroplasts in algae and plants.

The history of life was that of the unicellular eukaryotes, prokaryotes and archaea until about 610 million years ago when multicellular organisms began to appear in the oceans in the Ediacaran period. The evolution of multicellularity occurred in multiple independent events, in organisms as diverse as sponges, brown algae, cyanobacteria, slime moulds and myxobacteria. In January 2016, scientists reported that, about 800 million years ago, a minor genetic change in a single molecule called GK-PID may have allowed organisms to go from a single cell organism to one of many cells.

Soon after the emergence of these first multicellular organisms, a remarkable amount of biological diversity appeared over approximately 10 million years, in an event called the Cambrian explosion. Here, the majority of types of modern animals appeared in the fossil record, as well as unique lineages that subsequently became extinct. Various triggers for the Cambrian explosion have been proposed, including the accumulation of oxygen in the atmosphere from photosynthesis.

About 500 million years ago, plants and fungi colonised the land and were soon followed by arthropods and other animals. Insects were particularly successful and even today make up the majority of animal species. Amphibians first appeared around 364 million years ago, followed by early amniotes and birds around 155 million years ago (both from "reptile"-like lineages), mammals around 129 million years ago, homininae around 10 million years ago and modern humans around 250,000 years ago. However, despite the evolution of these large animals, smaller organisms similar to the types that evolved early in this process continue to be highly successful and dominate the Earth, with the majority of both biomass and species being prokaryotes.

Estimates on the number of Earth's current species range from 10 million to 14 million, of which about 1.2 million have been documented and over 86 percent have not yet been described.

Marine Microorganisms

Microbial Mats

Microbial mats are the earliest form of life on Earth for which there is good fossil evidence. The image shows a cyanobacterial-algal mat.

Stromatolites are formed from microbial mats as microbes slowly move upwards to avoid being smothered by sediment.

A microorganism (or microbe) is a microscopic living organism, which may be single-celled or multicellular. Microorganisms are very diverse and include all bacteria, archaea and most protozoa. This group also contains some species of fungi, algae, and certain microscopic animals, such as rotifers.

Many macroscopic animals and plants have microscopic juvenile stages. Some microbiologists also classify viruses (and viroids) as microorganisms, but others consider these as nonliving. In July 2016, scientists reported identifying a set of 355 genes from the last universal common ancestor (LUCA) of all life, including microorganisms, living on Earth.

Microorganisms live in every part of the biosphere, including soil, hot springs, "seven miles deep" in the ocean, "40 miles high" in the atmosphere and inside rocks far down within the Earth's crust. Microorganisms, under certain test conditions, have been observed to thrive in the vacuum of outer space. According to some estimates, microorganisms outweigh "all other living things combined thousands of times over". The mass of prokaryote microorganisms — which includes bacteria and archaea, but not the nucleated eukaryote microorganisms — may be as much as 0.8 trillion tons of carbon (of the total biosphere mass, estimated at between 1 and 4 trillion tons). On 17 March 2013, researchers reported data that suggested microbial life forms thrive in the Mariana Trench. the deepest spot in the Earth's oceans. Other researchers reported related studies that microorganisms thrive inside rocks up to 580 m (1,900 ft; 0.36 mi) below the sea floor under 2,590 m (8,500 ft; 1.61 mi) of ocean off the coast of the northwestern United States, as well as 2,400 m (7,900 ft; 1.5 mi) beneath the seabed off Japan. On 20 August 2014, scientists confirmed the existence of microorganisms living 800 m (2,600 ft; 0.50 mi) below the ice of Antarctica. According to one researcher, "You can find microbes everywhere — they're extremely adaptable to conditions, and survive wherever they are."

Microorganisms are crucial to nutrient recycling in ecosystems as they act as decomposers. A small proportion of microorganisms are pathogenic, causing disease and even death in plants and animals.

Marine Viruses

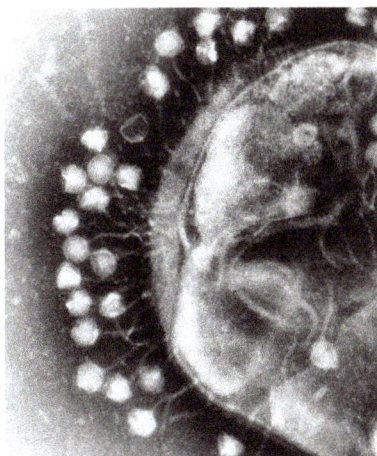

Transmission electron micrograph of multiple bacteriophages attached to a bacterial cell wall

A virus is a small infectious agent that replicates only inside the living cells of other organisms. Viruses can infect all types of life forms, from animals and plants to microorganisms, including bacteria and archaea.

When not inside an infected cell or in the process of infecting a cell, viruses exist in the form of independent particles. These viral particles, also known as *virions*, consist of two or three parts: (i) the genetic material made from either DNA or RNA, long molecules that carry genetic information; (ii) a protein coat, called the capsid, which surrounds and protects the genetic material; and in some cases (iii) an envelope of lipids that surrounds the protein coat when they are outside a cell. The shapes of these virus particles range from simple helical and icosahedral forms for some virus species to more complex structures for others. Most virus species have virions that are too small to be seen with an optical microscope. The average virion is about one one-hundredth the size of the average bacterium.

The origins of viruses in the evolutionary history of life are unclear: some may have evolved from plasmids—pieces of DNA that can move between cells—while others may have evolved from bacteria. In evolution, viruses are an important means of horizontal gene transfer, which increases genetic diversity. Viruses are considered by some to be a life form, because they carry genetic material, reproduce, and evolve through natural selection. However they lack key characteristics (such as cell structure) that are generally considered necessary to count as life. Because they possess some but not all such qualities, viruses have been described as "organisms at the edge of life" and as replicators.

Viruses are found wherever there is life and have probably existed since living cells first evolved. The origin of viruses is unclear because they do not form fossils, so molecular techniques have been used to compare the DNA or RNA of viruses and are a useful means of investigating how they arose.

Viruses are now recognised as ancient and as having origins that pre-date the divergence of life into the three domains.

Opinions differ on whether viruses are a form of life, or organic structures that interact with living organisms. They have been described as "organisms at the edge of life", since they resemble organisms in that they possess genes, evolve by natural selection, and reproduce by creating multiple copies of themselves through self-assembly. Although they have genes, they do not have a cellular structure, which is often seen as the basic unit of life. Viruses do not have their own metabolism, and require a host cell to make new products. They therefore cannot naturally reproduce outside a host cell.

Bacterial viruses, called bacteriophages, are a common and diverse group of viruses and are the most abundant form of biological entity in aquatic environments – there are up to ten times more of these viruses in the oceans than there are bacteria, reaching levels of 250,000,000 bacteriophages per millilitre of seawater.

There are also archaean viruses which replicate within archaea: these are double-stranded DNA viruses with unusual and sometimes unique shapes. These viruses have been studied in most detail in the thermophilic archaea, particularly the orders Sulfolobales and Thermoproteales.

A teaspoon of seawater contains about one million viruses. Most of these are bacteriophages, which are harmless to plants and animals, and are in fact essential to the regulation of saltwater and freshwater ecosystems. They infect and destroy bacteria in aquatic microbial communities, and are the most important mechanism of recycling carbon in the marine environment. The organic molecules released from the dead bacterial cells stimulate fresh bacterial and algal growth. Viral activity may also contribute to the biological pump, the process whereby carbon is sequestered in the deep ocean.

Microorganisms constitute more than 90% of the biomass in the sea. It is estimated that viruses kill approximately 20% of this biomass each day and that there are 15 times as many viruses in the oceans as there are bacteria and archaea. Viruses are the main agents responsible for the rapid destruction of harmful algal blooms, which often kill other marine life. The number of viruses in the oceans decreases further offshore and deeper into the water, where there are fewer host organisms.

Viruses are an important natural means of transferring genes between different species, which increases genetic diversity and drives evolution. It is thought that viruses played a central role in the early evolution, before the diversification of bacteria, archaea and eukaryotes, at the time of the last universal common ancestor of life on Earth. Viruses are still one of the largest reservoirs of unexplored genetic diversity on Earth.

Marine Bacteria

Pompeii worm able to survive temperatures as high as 176°F. A coating of protective bacteria covers this deep-sea worm's back.

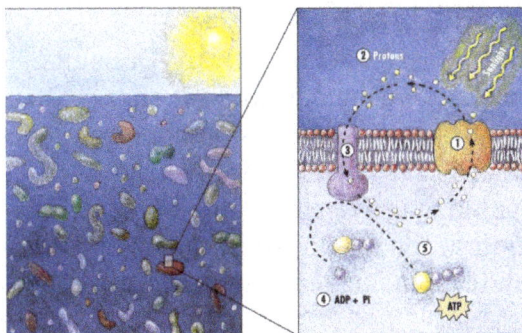

Energy gathering mechanism in marine bacteria via *Proteorhodopsin*

Bacteria constitute a large domain of prokaryotic microorganisms. Typically a few micrometres in length, bacteria have a number of shapes, ranging from spheres to rods and spirals. Bacteria were among the first life forms to appear on Earth, and are present in most of its habitats. Bacteria inhabit soil, water, acidic hot springs, radioactive waste, and the deep portions of Earth's crust. Bacteria also live in symbiotic and parasitic relationships with plants and animals.

Once regarded as plants constituting the class *Schizomycetes*, bacteria are now classified as prokaryotes. Unlike cells of animals and other eukaryotes, bacterial cells do not contain a nucleus and rarely harbour membrane-bound organelles. Although the term *bacteria* traditionally included all prokaryotes, the scientific classification changed after the discovery in the 1990s that prokaryotes consist of two very different groups of organisms that evolved from an ancient common ancestor. These evolutionary domains are called *Bacteria* and *Archaea*.

The ancestors of modern bacteria were unicellular microorganisms that were the first forms of life to appear on Earth, about 4 billion years ago. For about 3 billion years, most organisms were microscopic, and bacteria and archaea were the dominant forms of life. Although bacterial fossils exist, such as stromatolites, their lack of distinctive morphology prevents them from being used to examine the history of bacterial evolution, or to date the time of origin of a particular bacterial species. However, gene sequences can be used to reconstruct the bacterial phylogeny, and these studies indicate that bacteria diverged first from the archaeal/eukaryotic lineage. Bacteria were also involved in the second great evolutionary divergence, that of the archaea and eukaryotes. Here, eukaryotes resulted from the entering of ancient bacteria into endosymbiotic associations with the ancestors of eukaryotic cells, which were themselves possibly related to the Archaea. This involved the engulfment by proto-eukaryotic cells of alphaproteobacterial symbionts to form either mitochondria or hydrogenosomes, which are still found in all known Eukarya. Later on, some eukaryotes that already contained mitochondria also engulfed cyanobacterial-like organisms. This led to the formation of chloroplasts in algae and plants. There are also some algae that originated from even later endosymbiotic events. Here, eukaryotes engulfed a eukaryotic algae that developed into a "second-generation" plastid. This is known as secondary endosymbiosis.

In 2000 a team of microbiologists led by Edward DeLong made a crucial discovery in the understanding of the marine carbon and energy cycles. They discovered a gene in several species of bacteria responsible for production of the protein rhodopsin, previously unheard of in the domain Bacteria. These proteins found in the cell membranes are capable of converting light energy to biochemical energy due to a change in configuration of the rhodopsin molecule as sunlight strikes it, causing the pumping of a proton from inside out and a subsequent inflow that generates the energy.

Marine Archaea

Algae and archaea growing in the run-off of a geothermal geyser

The archaea constitute a domain and kingdom of single-celled microorganisms. These microbes are prokaryotes, meaning that they have no cell nucleus or any other membrane-bound organelles in their cells.

Archaea were initially classified as bacteria, but this classification is outdated. Archaeal cells have unique properties separating them from the other two domains of life, Bacteria and Eukaryota. The Archaea are further divided into multiple recognized phyla. Classification is difficult because the majority have not been isolated in the laboratory and have only been detected by analysis of their nucleic acids in samples from their environment.

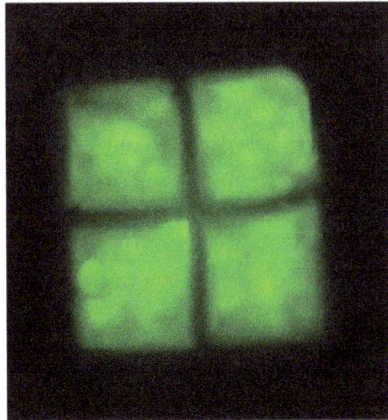

The flat and square-shaped cells of the archaea *Haloquadratum walsbyi*

Archaea and bacteria are generally similar in size and shape, although a few archaea have very strange shapes, such as the flat and square-shaped cells of *Haloquadratum walsbyi*. Despite this morphological similarity to bacteria, archaea possess genes and several metabolic pathways that are more closely related to those of eukaryotes, notably the enzymes involved in transcription and translation. Other aspects of archaeal biochemistry are unique, such as their reliance on ether lipids in their cell membranes, such as archaeols. Archaea use more energy sources than eukaryotes: these range from organic compounds, such as sugars, to ammonia, metal ions or even hydrogen gas. Salt-tolerant archaea (the Haloarchaea) use sunlight as an energy source, and other species of archaea fix carbon; however, unlike plants and cyanobacteria, no known species of archaea does both. Archaea reproduce asexually by binary fission, fragmentation, or budding; unlike bacteria and eukaryotes, no known species forms spores.

Archaea were initially viewed as extremophiles living in harsh environments, such as hot springs and salt lakes, but they have since been found in a broad range of habitats. Archaea are particularly numerous in the oceans, and the archaea in plankton may be one of the most abundant groups of organisms on the planet. Archaea are a major part of Earth's life and may play roles in both the carbon cycle and the nitrogen cycle.

Other Microorganisms

As inhabitants of the largest environment on Earth, microbial marine systems drive changes in every global system. Microbes are responsible for virtually all the photosynthesis that occurs in the ocean, as well as the cycling of carbon, nitrogen, phosphorus and other nutrients and trace elements.

Microscopic life undersea is incredibly diverse and still poorly understood. For example, the role of viruses in marine ecosystems is barely being explored even in the beginning of the 21st century.

A teaspoon of seawater contains about one million viruses. Most of these are bacteriophages, which are harmless to plants and animals, and are in fact essential to the regulation of saltwater and freshwater ecosystems. They infect and destroy bacteria in aquatic microbial communities, and are the most important mechanism of recycling carbon in the marine environment. The organic molecules released from the dead bacterial cells stimulate fresh bacterial and algal growth. Viral activity may also contribute to the biological pump, the process whereby carbon is sequestered in the deep ocean.

Marine bacteriophages are viruses that live as obligate parasitic agents in marine bacteria such as cyanobacteria. Their existence was discovered through electron microscopy and epifluorescence microscopy of ecological water samples, and later through metagenomic sampling of uncultured viral samples. The tailed bacteriophages appear to dominate marine ecosystems in number and diversity of organisms. However, viruses belonging to families Corticoviridae, Inoviridae and Microviridae are also known to infect diverse marine bacteria. Metagenomic evidence suggests that microviruses (icosahedral ssDNA phages) are particularly prevalent in marine habitats.

Bacteriophages, viruses that are parasitic on bacteria, were first discovered in the early twentieth century. Scientists today consider that their importance in ecosystems, particularly marine ecosystems, has been underestimated, leading to these infectious agents being poorly investigated and their numbers and species biodiversity being greatly under reported.

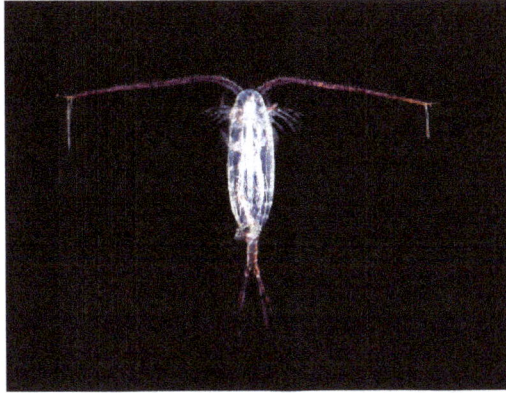
Copepod

Microorganisms constitute more than 90% of the biomass in the sea. It is estimated that viruses kill approximately 20% of this biomass each day and that there are 15 times as many viruses in the oceans as there are bacteria and archaea. Viruses are the main agents responsible for the rapid destruction of harmful algal blooms, which often kill other marine life. The number of viruses in the oceans decreases further offshore and deeper into the water, where there are fewer host organisms.

The role of phytoplankton is better understood due to their critical position as the most numerous primary producers on Earth. Phytoplankton are categorized into cyanobacteria (also called blue-green algae/bacteria), various types of algae (red, green, brown, and yellow-green), diatoms, dinoflagellates, euglenoids, coccolithophorids, cryptomonads, chrysophytes, chlorophytes, prasinophytes, and silicoflagellates.

Zooplankton tend to be somewhat larger, and not all are microscopic. Many Protozoa are zooplankton, including dinoflagellates, zooflagellates, foraminiferans, and radiolarians. Some of these (such as dinoflagellates) are also phytoplankton; the distinction between plants and animals often breaks down in very small organisms. Other zooplankton include cnidarians, ctenophores, chaetognaths, molluscs, arthropods, urochordates, and annelids such as polychaetes. Many larger animals begin their life as zooplankton before they become large enough to take their familiar forms. Two examples are fish larvae and sea stars (also called starfish).

Marine Algae and Plants

Microscopic algae and plants provide important habitats for life, sometimes acting as hiding and foraging places for larval forms of larger fish and invertebrates.

Algal life is widespread and very diverse under the ocean. Microscopic photosynthetic algae contribute a larger proportion of the world's photosynthetic output than all the terrestrial forests combined. Most of the niche occupied by sub plants on land is actually occupied by macroscopic algae in the ocean, such as *Sargassum* and kelp, which are commonly known as seaweeds that create kelp forests.

Plants that survive in the sea are often found in shallow waters, such as the seagrasses (examples of which are eelgrass, *Zostera*, and turtle grass, *Thalassia*). These plants have adapted to the high salinity of the ocean environment. The intertidal zone is also a good place to find plant life in the sea, where mangroves or cordgrass or beach grass might grow. Microscopic algae and plants provide important habitats for life, sometimes acting as hiding and foraging places for larval forms of larger fish and invertebrates.

Marine Invertebrates

Crown-of-thorns starfish

As on land, invertebrates make up a huge portion of all life in the sea. Invertebrate sea life includes Cnidaria such as jellyfish and sea anemones; Ctenophora; sea worms including the phyla Platyhelminthes, Nemertea, Annelida, Sipuncula, Echiura, Chaetognatha, and Phoronida; Mollusca including shellfish, squid, octopus; Arthropoda including Chelicerata and Crustacea; Porifera; Bryozoa; Echinodermata including starfish; and Urochordata including sea squirts or tunicates.

Marine Fungi

Over 1500 species of fungi are known from marine environments. These are parasitic on marine algae or animals, or are saprobes on algae, corals, protozoan cysts, sea grasses, wood and other substrata, and can also be found in sea foam. Spores of many species have special appendages which facilitate attachment to the substratum. A very diverse range of unusual secondary metabolites is produced by marine fungi.

Marine Vertebrates

Marine Fish

Fish anatomy includes a two-chambered heart, operculum, swim bladder, scales, eyes adapted to seeing underwater, and secretory cells that produce mucous. Fish breathe by extracting oxygen from water through gills. Fins propel and stabilize the fish in the water. Fish fall into two main groups: fish with bony skeletons and fish with cartilaginous skeletons.

A reported 32,700 species of fish have been described (as of December 2013), more than the combined total of all other vertebrates. About 60% of fish species are saltwater fish.

Marine Reptiles

Green turtle

Reptiles which inhabit or frequent the sea include sea turtles, sea snakes, terrapins, the marine iguana, and the saltwater crocodile. Most extant marine reptiles, except for some sea snakes, are oviparous and need to return to land to lay their eggs. Thus most species, excepting sea turtles, spend most of their lives on or near land rather than in the ocean. Despite their marine adaptations, most sea snakes prefer shallow waters nearby land, around islands, especially waters that are somewhat sheltered, as well as near estuaries. Some extinct marine reptiles, such as ichthyosaurs, evolved to be viviparous and had no requirement to return to land.

Marine Birds

Birds adapted to living in the marine environment are often called seabirds. Examples include albatross, penguins, gannets, and auks. Although they spend most of their lives in the ocean, species such as gulls can often be found thousands of miles inland.

Marine Mammals

There are five main types of marine mammals.

- Cetaceans include toothed whales (suborder Odontoceti), such as the sperm whale, dolphins, and porpoises such as the Dall's porpoise. Cetaceans also include baleen whales (suborder Mysticeti), such as the gray whale, humpback whale, and blue whale.

- Sirenians include manatees, the dugong, and the extinct Steller's sea cow.

- Seals (family Phocidae), sea lions (family Otariidae - which also include the fur seals), and the walrus (family Odobenidae) are all considered pinnipeds.

- The sea otter is a member of the family Mustelidae, which includes weasels and badgers.

- The polar bear is a member of the family Ursidae.

Plankton

Extinctions

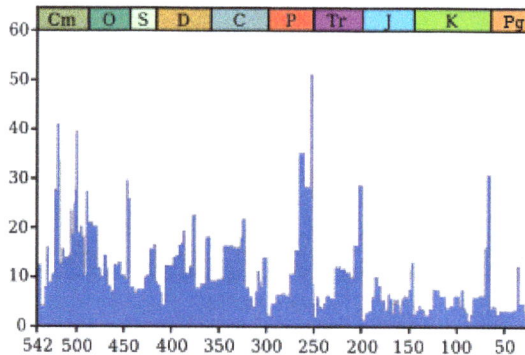

Apparent extinction intensity, i.e. the fraction of genera going extinct at any given time as reconstructed from the fossil record (excluding the current Holocene extinction event)

More than 99 percent of all species, amounting to over five billion species, that ever lived on Earth are estimated to be extinct. Extinction events occur when life undergoes precipitous global declines. Most diversity and biomass on earth is found among the microorganisms, which are difficult to measure. Recorded extinction events are therefore based on the more easily observed changes in the diversity and abundance of larger multicellular organisms, rather than the total diversity and abundance of life.

Extinction occurs at an uneven rate. Based on the fossil record, the background rate of extinctions on Earth is about two to five taxonomic families of marine animals every million years. Marine fossils are mostly used to measure extinction rates because of their superior fossil record and stratigraphic range compared to land organisms.

The Great Oxygenation Event was perhaps the first major extinction event. Since the Cambrian explosion five further major mass extinctions have significantly exceeded the background extinction rate. In addition to these major mass extinctions there are numerous minor ones, as well as the current ongoing mass-extinction caused by human activity, the Holocene extinction sometimes called the "sixth extinction".

Marine Habitats

The marine environment supplies many kinds of habitats that support marine life. Marine life depends in some way on the saltwater that is in the sea (the term *marine* comes from the Latin *mare*, meaning sea or ocean). A habitat is an ecological or environmental area inhabited by one or more living species.

Marine habitats can be divided into coastal and open ocean habitats. Coastal habitats are found in the area that extends from as far as the tide comes in on the shoreline out to the edge of the continental shelf. Most marine life is found in coastal habitats, even though the shelf area occupies only seven percent of the total ocean area. Open ocean habitats are found in the deep ocean beyond the edge of the continental shelf.

Alternatively, marine habitats can be divided into pelagic and demersal zones. Pelagic habitats are found near the surface or in the open water column, away from the bottom of the ocean. Demersal habitats are near or on the bottom of the ocean. An organism living in a pelagic habitat is said to be a pelagic organism, as in pelagic fish. Similarly, an organism living in a demersal habitat is said to be a demersal organism, as in demersal fish. Pelagic habitats are intrinsically shifting and ephemeral, depending on what ocean currents are doing.

Marine habitats can be modified by their inhabitants. Some marine organisms, like corals, kelp, mangroves and seagrasses, are ecosystem engineers which reshape the marine environment to the point where they create further habitat for other organisms.

Overview

Only 29 percent of the world surface is land. The rest is ocean, home to the marine habitats. The oceans are nearly four kilometres deep on average and are fringed with coastlines that run for nearly 380,000 kilometres.

In contrast to terrestrial habitats, marine habitats are shifting and ephemeral. Swimming organisms find areas by the edge of a continental shelf a good habitat, but only while upwellings bring nutrient rich water to the surface. Shellfish find habitat on sandy beaches, but storms, tides and currents mean their habitat continually reinvents itself.

The presence of seawater is common to all marine habitats. Beyond that many other things determine whether a marine area makes a good habitat and the type of habitat it makes. For example:

- temperature – is affected by geographical latitude, ocean currents, weather, the discharge of rivers, and by the presence of hydrothermal vents or cold seeps

- sunlight – photosynthetic processes depend on how deep and turbid the water is

- nutrients – are transported by ocean currents to different marine habitats from land runoff, or by upwellings from the deep sea, or they sink though the sea as marine snow

- salinity – varies, particularly in estuaries or near river deltas, or by hydrothermal vents

- dissolved gases – oxygen levels in particular, can be increased by wave actions and decreased during algal blooms

- acidity – this is partly to do with dissolved gases above, since the acidity of the ocean is largely controlled by how much carbon dioxide is in the water.

- turbulence – ocean waves, fast currents and the agitation of water affect the nature of habitats

- cover – the availability of cover such as the adjacency of the sea bottom, or the presence of floating objects

- the occupying organisms themselves – since organisms modify their habitats by the act of occupying them, and some, like corals, kelp, mangroves and seagrasses, create further habitats for other organisms.

Ocean	Area million km²	%	Volume million cu km	%	Mean depth km	Max depth km	Coast-line km	%	Ref
Pacific Ocean	155.6	46.4	679.6	49.6	4.37	10.924	135,663		
Atlantic Ocean	76.8	22.9	313.4	22.5	4.08	8.605	111,866		
Indian Ocean	68.6	20.4	269.3	19.6	3.93	7.258	66,526		
Southern Ocean	20.3	6.1	91.5	6.7	4.51	7.235	17,968		
Arctic Ocean	14.1	4.2	17.0	1.2	1.21	4.665	45,389		
Overall	335.3		1370.8		4.09	10.924	377,412		

The ocean occupies 71 percent of the world surface, averaging nearly four kilometres in depth. There are five major oceans, of which the Pacific Ocean is nearly as large as the rest put together. Coastlines fringe the land for nearly 380,000 kilometres.

Land runoff, pouring into the sea, can contain nutrients

Marine habitats can be broadly divided into pelagic and demersal habitats. Pelagic habitats are the habitats of the open water column, away from the bottom of the ocean. Demersal habitats are the habitats that are near or on the bottom of the ocean. An organism living in a pelagic habitat is said to be a pelagic organism, as in pelagic fish. Similarly, an organism living in a demersal habitat is said to be a demersal organism, as in demersal fish. Pelagic habitats are intrinsically ephemeral, depending on what ocean currents are doing.

The land-based ecosystem depends on topsoil and fresh water, while the marine ecosystem depends on dissolved nutrients washed down from the land.

Ocean deoxygenation poses a threat to marine habitats, due to the growth of low oxygen zones.

Ocean Currents

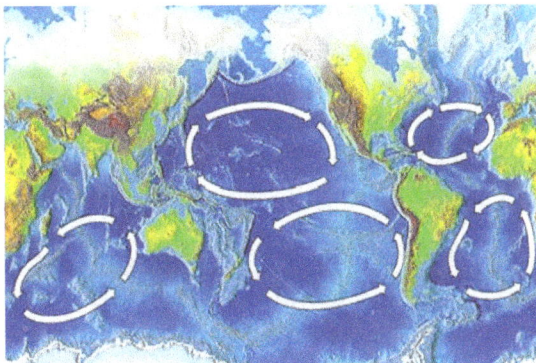
Ocean gyres rotate clockwise in the north and counterclockwise in the south

In marine systems, ocean currents have a key role determining which areas are effective as habitats, since ocean currents transport the basic nutrients needed to support marine life. Plankton are the life forms that inhabit the ocean that are so small (less than 2 mm) that they cannot effectively propel themselves through the water, but must drift instead with the currents. If the current carries the right nutrients, and if it also flows at a suitably shallow depth where there is plenty of sunlight, then such a current itself can become a suitable habitat for photosynthesizing tiny algae called phytoplankton. These tiny plants are the primary producers in the ocean, at the start of the food chain. In turn, as the population of drifting phytoplankton grows, the water becomes a suitable habitat for zooplankton, which feed on the phytoplankton. While phytoplankton are tiny drifting plants, zooplankton are tiny drifting animals, such as the larvae of fish and marine invertebrates. If sufficient zooplankton establish themselves, the current becomes a candidate habitat for the forage fish that feed on them. And then if sufficient forage fish move to the area, it becomes a candidate habitat for larger predatory fish and other marine animals that feed on the forage fish. In this dynamic way, the current itself can, over time, become a moving habitat for multiple types of marine life.

This algae bloom occupies sunlit epipelagic waters off the southern coast of England. The algae are maybe feeding on nutrients from land runoff or upwellings at the edge of the continental shelf

Ocean currents can be generated by differences in the density of the water. How dense water is depends on how saline or warm it is. If water contains differences in salt content or temperature, then the different densities will initiate a current. Water that is saltier or cooler will be denser, and will sink in relation to the surrounding water. Conversely, warmer and less salty water will float to the surface. Atmospheric winds and pressure differences also produces surface currents, waves and seiches. Ocean currents are also generated by the gravitational pull of the sun and moon (tides), and seismic activity (tsunami).

The rotation of the Earth affects the direction ocean currents take, and explains which way the large circular ocean gyres rotate in the image above left. Suppose a current at the equator is heading north. The Earth rotates eastward, so the water possesses that rotational momentum. But the further the water moves north, the slower the earth moves eastward. If the current could get to the North Pole, the earth wouldn't be moving eastward at all. To conserve its rotational momentum, the further the current travels north the faster it must move eastward. So the effect is that the current curves to the right. This is the Coriolis effect. It is weakest at the equator and strongest at the poles. The effect is opposite south of the equator, where currents curve left.

Marine Topography

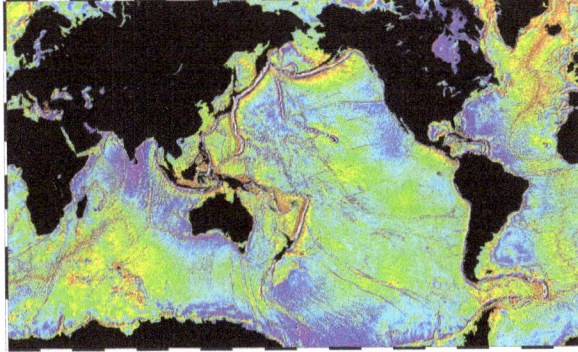

Map of underwater topography (1995 NOAA)

Marine topography refers to the shape the land has when it interfaces with the ocean. These shapes are obvious along coastlines, but they occur also in significant ways underwater. The effectiveness of marine habitats is partially defined by these shapes, including the way they interact with and shape ocean currents, and the way sunlight diminishes when these landforms occupy increasing depths.

Marine topographies include coastal and oceanic landforms ranging from coastal estuaries and shorelines to continental shelves and coral reefs. Further out in the open ocean, they include underwater and deep sea features such as ocean rises and seamounts. The submerged surface has mountainous features, including a globe-spanning mid-ocean ridge system, as well as undersea volcanoes, oceanic trenches, submarine canyons, oceanic plateaus and abyssal plains.

The mass of the oceans is approximately 1.35×10^{18} metric tons, or about $1/4400$ of the total mass of the Earth. The oceans cover an area of 3.618×10^8 km² with a mean depth of 3,682 m, resulting in an estimated volume of 1.332×10^9 km³.

Biomass

One measure of the relative importance of different marine habitats is the rate at which they produce biomass.

Producer	Biomass productivity (gC/m²/yr)	Ref	Total area (million km²)	Ref	Total production (billion tonnes C/yr)	Comment
swamps and marshes	2,500					Includes freshwater

Producer	Biomass productivity (gC/m²/ yr)	Ref	Total area (million km²)	Ref	Total production (billion tonnes C/yr)	Comment
coral reefs	2,000		0.28		0.56	
algal beds	2,000					
river estuaries	1,800					
open ocean	125		311		39	

Coastal

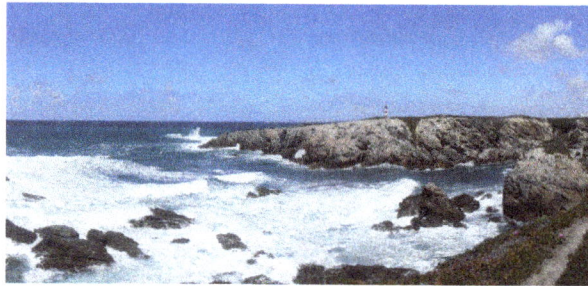
Coastlines can be volatile habitats

Marine coasts are dynamic environments which constantly change, like the ocean which partially shape them. The Earth's natural processes, including weather and sea level change, result in the erosion, accretion and resculpturing of coasts as well as the flooding and creation of continental shelves and drowned river valleys.

The main agents responsible for deposition and erosion along coastlines are waves, tides and currents. The formation of coasts also depends on the nature of the rocks they are made of – the harder the rocks the less likely they are to erode, so variations in rock hardness result in coastlines with different shapes.

Tides often determine the range over which sediment is deposited or eroded. Areas with high tidal ranges allow waves to reach farther up the shore, and areas with lower tidal ranges produce deposition at a smaller elevation interval. The tidal range is influenced by the size and shape of the coastline. Tides do not typically cause erosion by themselves; however, tidal bores can erode as the waves surge up river estuaries from the ocean.

Waves erode coastline as they break on shore releasing their energy; the larger the wave the more energy it releases and the more sediment it moves. Sediment deposited by waves comes from eroded cliff faces and is moved along the coastline by the waves. Sediment deposited by rivers is the dominant influence on the amount of sediment located on a coastline.

Shores that look permanent through the short perceptive of a human lifetime are in fact among the most temporary of all marine structures.

The sedimentologist Francis Shepard classified coasts as *primary* or *secondary*.

- Primary coasts are shaped by non-marine processes, by changes in the land form. If a coast is in much the same condition as it was when sea level was stabilised after the last ice age, it is called a primary coast. "Primary coasts are created by erosion (the wearing away of soil or rock), deposition (the buildup of sediment or sand) or tectonic activity (changes in the structure of the rock and soil because of earthquakes). Many of these coastlines were formed as the sea level rose during the last 18,000 years, submerging river and glacial valleys to form bays and fjords." An example of a primary coast is a river delta, which forms when a river deposits soil and other material as it enters the sea.

- Secondary coasts are produced by marine processes, such as the action of the sea or by creatures that live in it. Secondary coastlines include sea cliffs, barrier islands, mud flats, coral reefs, mangrove swamps and salt marshes.

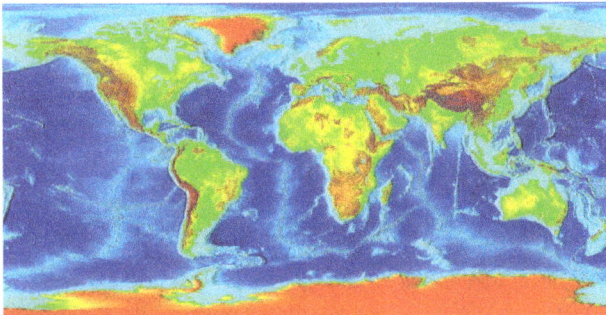

The global continental shelf, highlighted in cyan, defines the extent of coastal habitats, and occupies 5% of the total world area.

Continental coastlines usually have a continental shelf, a shelf of relatively shallow water, less than 200 metres deep, which extends 68 km on average beyond the coast. Worldwide, continental shelves occupy a total area of about 24 million km² (9 million sq mi), 8% of the ocean's total area and nearly 5% of the world's total area. Since the continental shelf is usually less than 200 metres deep, it follows that coastal habitats are generally photic, situated in the sunlit epipelagic zone. This means the conditions for photosynthetic processes so important for primary production, are available to coastal marine habitats. Because land is nearby, there are large discharges of nutrient rich land runoff into coastal waters. Further, periodic upwellings from the deep ocean can provide cool and nutrient rich currents along the edge of the continental shelf.

As a result, coastal marine life is the most abundant in the world. It is found in tidal pools, fjords and estuaries, near sandy shores and rocky coastlines, around coral reefs and on or above the continental shelf. Coastal fish include small forage fish as well as the larger predator fish that feed on them. Forage fish thrive in inshore waters where high productivity results from upwelling and shoreline run off of nutrients. Some are partial residents that spawn in streams, estuaries and bays, but most complete their life cycle in the zone. There can also be a mutualism between species that occupy adjacent marine habitats. For example, fringing reefs just below low tide level have a mutually beneficial relationship with mangrove forests at high tide level and sea grass meadows in between: the reefs protect the mangroves and seagrass from strong currents and waves that would damage them or erode the sediments in which they are rooted, while the mangroves and seagrass protect the coral from large influxes of silt, fresh water and pollutants. This additional level of variety in the environment is beneficial to many types of coral reef animals, which for example may feed in the sea grass and use the reefs for protection or breeding.

Coastal habitats are the most visible marine habitats, but they are not the only important marine habitats. Coastlines run for 380,000 kilometres, and the total volume of the ocean is 1,370 million cu km. This means that for each metre of coast, there is 3.6 cu km of ocean space available somewhere for marine habitats.

Waves and currents shape the intertidal shoreline, eroding the softer rocks and transporting and grading loose particles into shingles, sand or mud

Intertidal

Intertidal zones, those areas close to shore, are constantly being exposed and covered by the ocean's tides. A huge array of life lives within this zone.

Shore habitats range from the upper intertidal zones to the area where land vegetation takes prominence. It can be underwater anywhere from daily to very infrequently. Many species here are scavengers, living off of sea life that is washed up on the shore. Many land animals also make much use of the shore and intertidal habitats. A sub-group of organisms in this habitat bores and grinds exposed rock through the process of bioerosion.

Sandy Shores

Sandy shores provide shifting homes to many species

Sandy shores, also called beaches, are coastal shorelines where sand accumulates. Waves and currents shift the sand, continually building and eroding the shoreline. Longshore currents flow parallel to the beaches, making waves break obliquely on the sand. These currents transport large amounts of sand along coasts, forming spits, barrier islands and tombolos. Longshore currents also commonly create offshore bars, which give beaches some stability by reducing erosion.

Sandy shores are full of life, The grains of sand host diatoms, bacteria and other microscopic creatures. Some fish and turtles return to certain beaches and spawn eggs in the sand. Birds habitat beaches, like gulls, loons, sandpipers, terns and pelicans. Aquatic mammals, such sea lions, recuperate on them. Clams, periwinkles, crabs, shrimp, starfish and sea urchins are found on most beaches.

Sand is a sediment made from small grains or particles with diameters between about 60 μm and 2 mm. Mud is a sediment made from particles finer than sand. This small particle size means that mud particles tend to stick together, whereas sand particles do not. Mud is not easily shifted by waves and currents, and when it dries out, cakes into a solid. By contrast, sand is easily shifted by waves and currents, and when sand dries out it can be blown in the wind, accumulating into shifting sand dunes. Beyond the high tide mark, if the beach is low-lying, the wind can form rolling hills of sand dunes. Small dunes shift and reshape under the influence of the wind while larger dunes stabilise the sand with vegetation.

Ocean processes grade loose sediments to particle sizes other than sand, such as gravel or cobbles. Waves breaking on a beach can leave a berm, which is a raised ridge of coarser pebbles or sand, at the high tide mark. Shingle beaches are made of particles larger than sand, such as cobbles, or small stones. These beaches make poor habitats. Little life survives because the stones are churned and pounded together by waves and currents.

Rocky Shores

The relative solidity of rocky shores seems to give them a permanence compared to the shifting nature of sandy shores. This apparent stability is not real over even quite short geological time scales, but it is real enough over the short life of an organism. In contrast to sandy shores, plants and animals can anchor themselves to the rocks.

Tidepools on rocky shores make turbulent habitats for many forms of marine life

Competition can develop for the rocky spaces. For example, barnacles can compete successfully on open intertidal rock faces to the point where the rock surface is covered with them. Barnacles resist desiccation and grip well to exposed rock faces. However, in the crevices of the same rocks, the inhabitants are different. Here mussels can be the successful species, secured to the rock with their byssal threads.

Rocky and sandy coasts are vulnerable because humans find them attractive and want to live near them. An increasing proportion of the humans live by the coast, putting pressure on coastal habitats.

Mudflats

Mudflats become temporary habitats for migrating birds

Mudflats are coastal wetlands that form when mud is deposited by tides or rivers. They are found in sheltered areas such as bays, bayous, lagoons, and estuaries. Mudflats may be viewed geologically as exposed layers of bay mud, resulting from deposition of estuarine silts, clays and marine animal detritus. Most of the sediment within a mudflat is within the intertidal zone, and thus the flat is submerged and exposed approximately twice daily.

Mudflats are typically important regions for wildlife, supporting a large population, although levels of biodiversity are not particularly high. They are of particular importance to migratory birds. In the United Kingdom mudflats have been classified as a Biodiversity Action Plan priority habitat.

Mangrove and Salt Marshes

Mangroves provide nurseries for fish

Mangrove swamps and salt marshes form important coastal habitats in tropical and temperate areas respectively.

Mangroves are species of shrubs and medium size trees that grow in saline coastal sediment habitats in the tropics and subtropics – mainly between latitudes 25° N and 25° S. The saline conditions tolerated by various species range from brackish water, through pure seawater (30 to 40 ppt), to water concentrated by evaporation to over twice the salinity of ocean seawater (up to 90 ppt). There are many mangrove species, not all closely related. The term "mangrove" is used generally to cover all of these species, and it can be used narrowly to cover just mangrove trees of the genus *Rhizophora*.

Mangroves form a distinct characteristic saline woodland or shrubland habitat, called a *mangrove swamp* or *mangrove forest*. Mangrove swamps are found in depositional coastal environments, where fine sediments (often with high organic content) collect in areas protected from high-energy wave action. Mangroves dominate three quarters of tropical coastlines.

Estuaries

An estuary is a partly enclosed coastal body of water with one or more rivers or streams flowing into it, and with a free connection to the open sea. Estuaries form a transition zone between river environments and ocean environments and are subject to both marine influences, such as tides, waves, and the influx of saline water; and riverine influences, such as flows of fresh water and sediment. The inflow of both seawater and freshwater provide high levels of nutrients in both the water column and sediment, making estuaries among the most productive natural habitats in the world.

Estuaries occur when rivers flow into a coastal bay or inlet. They are nutrient rich and have a transition zone which moves from freshwater to saltwater.

Most estuaries were formed by the flooding of river-eroded or glacially scoured valleys when sea level began to rise about 10,000-12,000 years ago. They are amongst the most heavily populated areas throughout the world, with about 60% of the world's population living along estuaries and the coast. As a result, estuaries are suffering degradation by many factors, including sedimentation from soil erosion from deforestation; overgrazing and other poor farming practices; overfishing; drainage and filling of wetlands; eutrophication due to excessive nutrients from sewage and animal wastes; pollutants including heavy metals, PCBs, radionuclides and hydrocarbons from sewage inputs; and diking or damming for flood control or water diversion.

Estuaries provide habitats for a large number of organisms and support very high productivity. Estuaries provide habitats for salmon and sea trout nurseries, as well as migratory bird populations. Two of the main characteristics of estuarine life are the variability in salinity and sedimentation. Many species of fish and invertebrates have various methods to control or conform to the shifts in salt concentrations and are termed osmoconformers and osmoregulators. Many animals also burrow to avoid predation and to live in the more stable sedimental environment. However, large numbers of bacteria are found within the sediment which have a very high oxygen demand. This reduces the levels of oxygen within the sediment often resulting in partially anoxic conditions, which can be further exacerbated by limited water flux. Phytoplankton are key primary producers in estuaries. They move with the water bodies and can be flushed in and out with the tides. Their productivity is largely dependent on the turbidity of the water. The main phytoplankton present are diatoms and dinoflagellates which are abundant in the sediment.

Kelp Forests

Kelp forests are underwater areas with a high density of kelp. They form some of the most productive and dynamic ecosystems on Earth. Smaller areas of anchored kelp are called *kelp beds*. Kelp forests occur worldwide throughout temperate and polar coastal oceans.

Kelp forests provide habitat for many marine organisms

Kelp forests provide a unique three-dimensional habitat for marine organisms and are a source for understanding many ecological processes. Over the last century, they have been the focus of extensive research, particularly in trophic ecology, and continue to provoke important ideas that are relevant beyond this unique ecosystem. For example, kelp forests can influence coastal oceanographic patterns and provide many ecosystem services.

However, humans have contributed to kelp forest degradation. Of particular concern are the effects of overfishing nearshore ecosystems, which can release herbivores from their normal population regulation and result in the over-grazing of kelp and other algae. This can rapidly result in transitions to barren landscapes where relatively few species persist.

Frequently considered an ecosystem engineer, kelp provides a physical substrate and habitat for kelp forest communities. In algae (Kingdom: Protista), the body of an individual organism is known as a thallus rather than as a plant (Kingdom: Plantae). The morphological structure of a kelp thallus is defined by three basic structural units:

- The holdfast is a root-like mass that anchors the thallus to the sea floor, though unlike true roots it is not responsible for absorbing and delivering nutrients to the rest of the thallus;

- The stipe is analogous to a plant stalk, extending vertically from the holdfast and providing a support framework for other morphological features;

- The fronds are leaf- or blade-like attachments extending from the stipe, sometimes along its full length, and are the sites of nutrient uptake and photosynthetic activity.

In addition, many kelp species have pneumatocysts, or gas-filled bladders, usually located at the base of fronds near the stipe. These structures provide the necessary buoyancy for kelp to maintain an upright position in the water column.

The environmental factors necessary for kelp to survive include hard substrate (usually rock), high nutrients (e.g., nitrogen, phosphorus), and light (minimum annual irradiance dose > 50 E m^{-2}). Especially productive kelp forests tend to be associated with areas of significant oceanographic upwelling, a process that delivers cool nutrient-rich water from depth to the ocean's mixed surface layer. Water flow and turbulence facilitate nutrient assimilation across kelp fronds throughout the water column. Water clarity affects the depth to which sufficient light can be transmitted. In ideal conditions, giant kelp (*Macrocystis spp.*) can grow as much as 30-60 centimetres vertically per day. Some species such as *Nereocystis* are annual while others like *Eisenia* are perennial, living for more than 20 years. In perennial kelp forests, maximum growth rates occur during upwelling months (typically spring and summer) and die-backs correspond to reduced nutrient availability, shorter photoperiods and increased storm frequency.

Seagrass Meadows

White-spotted puffers like living in seagrass areas

Seagrasses are flowering plants from one of four plant families which grow in marine environments. They are called *seagrasses* because the leaves are long and narrow and are very often green, and because the plants often grow in large meadows which look like grassland. Since seagrasses photosynthesize and are submerged, they must grow submerged in the photic zone, where there is enough sunlight. For this reason, most occur in shallow and sheltered coastal waters anchored in sand or mud bottoms.

Seagrasses form extensive beds or meadows, which can be either monospecific (made up of one species) or multispecific (where more than one species co-exist). Seagrass beds make highly diverse and productive ecosystems. They are home to phyla such as juvenile and adult fish, epiphytic and free-living macroalgae and microalgae, mollusks, bristle worms, and nematodes. Few species were originally considered to feed directly on seagrass leaves (partly because of their low nutritional content), but scientific reviews and improved working methods have shown that seagrass herbivory is a highly important link in the food chain, with hundreds of species feeding on seagrasses worldwide, including green turtles, dugongs, manatees, fish, geese, swans, sea urchins and crabs.

Seagrasses are ecosystem engineers in the sense that they partly create their own habitat. The leaves slow down water-currents increasing sedimentation, and the seagrass roots and rhizomes stabilize the seabed. Their importance to associated species is mainly due to provision of shelter (through their three-dimensional structure in the water column), and due to their extraordinarily high rate of primary production. As a result, seagrasses provide coastal zones with ecosystem services, such as fishing grounds, wave protection, oxygen production and protection against coastal erosion. Seagrass meadows account for 15% of the ocean's total carbon storage.

Coral Reefs

Reefs comprise some of the densest and most diverse habitats in the world. The best-known types of reefs are tropical coral reefs which exist in most tropical waters; however, reefs can also exist in cold water. Reefs are built up by corals and other calcium-depositing animals, usually on top of a rocky outcrop on the ocean floor. Reefs can also grow on other surfaces, which has made it possible to create artificial reefs. Coral reefs also support a huge community of life, including the corals themselves, their symbiotic zooxanthellae, tropical fish and many other organisms.

Much attention in marine biology is focused on coral reefs and the El Niño weather phenomenon. In 1998, coral reefs experienced the most severe mass bleaching events on record, when vast expanses of reefs across the world died because sea surface temperatures rose well above normal. Some reefs are recovering, but scientists say that between 50% and 70% of the world's coral reefs are now endangered and predict that global warming could exacerbate this trend.

Open Ocean

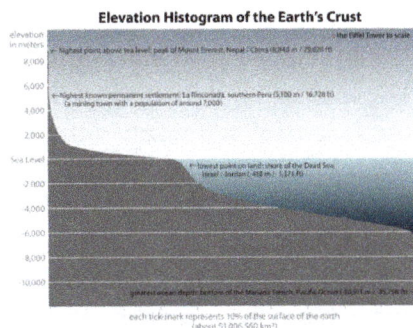

Elevation-area graph showing the proportion of land area at given heights and the proportion of ocean area at given depths

The open ocean is relatively unproductive because of a lack of nutrients, yet because it is so vast, it has more overall primary production than any other marine habitat. Only about 10 percent of marine species live in the open ocean. But among them are the largest and fastest of all marine animals, as well as the animals that dive the deepest and migrate the longest. In the depths lurk animal that, to our eyes, appear hugely alien.

Surface Waters

In the open ocean, sunlit surface epipelagic waters get enough light for photosynthesis, but there are often not enough nutrients. As a result, large areas contain little life apart from migrating animals.

The surface waters are sunlit. The waters down to about 200 metres are said to be in the epipelagic zone. Enough sunlight enters the epipelagic zone to allow photosynthesis by phytoplankton. The epipelagic zone is usually low in nutrients. This partially because the organic debris produced in the zone, such as excrement and dead animals, sink to the depths and are lost to the upper zone. Photosynthesis can happen only if both sunlight and nutrients are present.

In some places, like at the edge of continental shelves, nutrients can upwell from the ocean depth, or land runoff can be distributed by storms and ocean currents. In these areas, given that both sunlight and nutrients are now present, phytoplankton can rapidly establish itself, multiplying so fast that the water turns green from the chlorophyll, resulting in an algal bloom. These nutrient rich surface waters are among the most biologically productive in the world, supporting billions of tonnes of biomass.

"Phytoplankton are eaten by zooplankton - small animals which, like phytoplankton, drift in the ocean currents. The most abundant zooplankton species are copepods and krill: tiny crustaceans that are the most numerous animals on Earth. Other types of zooplankton include jelly fish and the larvae of fish, marine worms, starfish, and other marine organisms". In turn, the zooplankton are eaten by filter feeding animals, including some seabirds, small forage fish like herrings and sardines, whale sharks, manta rays, and the largest animal in the world, the blue whale. Yet again, moving up the foodchain, the small forage fish are in turn eaten by larger predators, such as tuna, marlin, sharks, large squid, seabirds, dolphins, and toothed whales.

Deep Sea

The deep sea starts at the aphotic zone, the point where sunlight loses most of its energy in the water. Many life forms that live at these depths have the ability to create their own light a unique evolution known as bio-luminescence.

In the deep ocean, the waters extend far below the epipelagic zone, and support very different types of pelagic life forms adapted to living in these deeper zones.

Much of the aphotic zone's energy is supplied by the open ocean in the form of detritus. In deep water, marine snow is a continuous shower of mostly organic detritus falling from the upper layers of the water column. Its origin lies in activities within the productive photic zone. Marine snow includes dead or dying plankton, protists (diatoms), fecal matter, sand, soot and other inorganic dust. The "snowflakes" grow over time and may reach several centimetres in diameter, travelling for weeks before reaching the ocean floor. However, most organic components of marine snow are consumed by microbes, zooplankton and other filter-feeding animals within the first 1,000 metres of their journey, that is, within the epipelagic zone. In this way marine snow may be considered the foundation of deep-sea mesopelagic and benthic ecosystems: As sunlight cannot reach them, deep-sea organisms rely heavily on marine snow as an energy source.

Some deep-sea pelagic groups, such as the lanternfish, ridgehead, marine hatchetfish, and lightfish families are sometimes termed *pseudoceanic* because, rather than having an even distribution in open water, they occur in significantly higher abundances around structural oases, notably seamounts and over continental slopes. The phenomenon is explained by the likewise abundance of prey species which are also attracted to the structures.

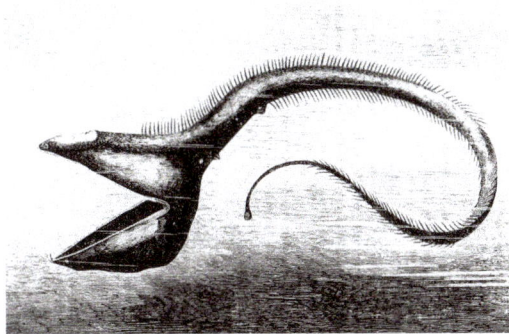

The umbrella mouth gulper eel can swallow a fish much larger than itself

The fish in the different pelagic and deep water benthic zones are physically structured, and behave in ways, that differ markedly from each other. Groups of coexisting species within each zone all seem to operate in similar ways, such as the small mesopelagic vertically migrating plankton-feeders, the bathypelagic anglerfishes, and the deep water benthic rattails. "

Ray finned species, with spiny fins, are rare among deep sea fishes, which suggests that deep sea fish are ancient and so well adapted to their environment that invasions by more modern fishes have been unsuccessful. The few ray fins that do exist are mainly in the Beryciformes and Lampriformes, which are also ancient forms. Most deep sea pelagic fishes belong to their own orders, suggesting a long evolution in deep sea environments. In contrast, deep water benthic species, are in orders that include many related shallow water fishes.

The umbrella mouth gulper is a deep sea eel with an enormous loosely hinged mouth. It can open its mouth wide enough to swallow a fish much larger than itself, and then expand its stomach to accommodate its catch.

Sea Floor

Vents and Seeps

Hydrothermal vents along the mid-ocean ridge spreading centers act as oases, as do their opposites, cold seeps. Such places support unique biomes and many new microbes and other lifeforms have been discovered at these locations.

Zooarium chimney provides a habitat for vent biota

Trenches

The deepest recorded oceanic trenches measure to date is the Mariana Trench, near the Philippines, in the Pacific Ocean at 10,924 m (35,838 ft). At such depths, water pressure is extreme and there is no sunlight, but some life still exists. A white flatfish, a shrimp and a jellyfish were seen by the American crew of the bathyscaphe *Trieste* when it dove to the bottom in 1960.

Seamounts

Marine life also flourishes around seamounts that rise from the depths, where fish and other sea life congregate to spawn and feed.

Marine Microorganism

A microorganism (or microbe) is a microscopic living organism, which may be single-celled or multicellular. Microorganisms are very diverse and include all bacteria, archaea and most protozoa. This group also contains some species of fungi, algae, and certain microscopic animals, such as rotifers.

Many macroscopic animals and plants have microscopic juvenile stages. Some microbiologists also classify viruses (and viroids) as microorganisms, but others consider these as nonliving. In July 2016, scientists reported identifying a set of 355 genes from the last universal common ancestor (LUCA) of all life, including microorganisms, living on Earth.

Microorganisms are crucial to nutrient recycling in ecosystems as they act as decomposers. A small proportion of microorganisms are pathogenic, causing disease and even death in plants and animals. As inhabitants of the largest environment on Earth, microbial marine systems drive changes in every global system. Microbes are responsible for virtually all the photosynthesis that occurs in the ocean, as well as the cycling of carbon, nitrogen, phosphorus and other nutrients and trace elements.

Microscopic life in the oceans is very diverse and still poorly understood. For example, the role of viruses in marine ecosystems has barely being explored even in the beginning of the 21st century.

Overview

microbial Mats

A teaspoon of seawater contains about one million viruses. Most of these are bacteriophages, which are harmless to plants and animals, and are in fact essential to the regulation of saltwater and freshwater ecosystems. They infect and destroy bacteria in aquatic microbial communities, and are the most important mechanism of recycling carbon in the marine environment. The organic molecules released from the dead bacterial cells stimulate fresh bacterial and algal growth. Viral activity may also contribute to the biological pump, the process whereby carbon is sequestered in the deep ocean.

Marine bacteriophages are viruses that live as obligate parasitic agents in marine bacteria such as cyanobacteria. Their existence was discovered through electron microscopy and epifluorescence microscopy of ecological water samples, and later through metagenomic sampling of uncultured viral samples. The tailed bacteriophages appear to dominate marine ecosystems in number and diversity of organisms. However, viruses belonging to families Corticoviridae, Inoviridae and Microviridae are also known to infect diverse marine bacteria. Metagenomic evidence suggests that microviruses (icosahedral ssDNA phages) are particularly prevalent in marine habitats.

Microbial mats are the earliest form of life on Earth for which there is good fossil evidence. The image shows a cyanobacterial-algal mat.

Stromatolites are formed from microbial mats as microbes slowly move upwards to avoid being smothered by sediment.

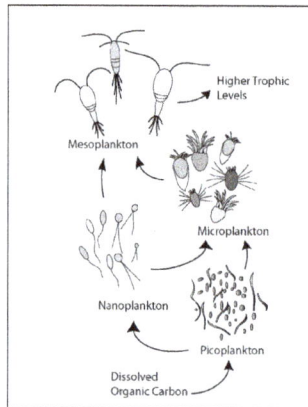

Marine microbial loop

Bacteriophages, viruses that are parasitic on bacteria, were first discovered in the early twentieth century. Scientists today consider that their importance in ecosystems, particularly marine ecosystems, has been underestimated, leading to these infectious agents being poorly investigated and their numbers and species biodiversity being greatly under reported.

Microorganisms constitute more than 90% of the biomass in the sea. It is estimated that viruses kill approximately 20% of this biomass each day and that there are 15 times as many viruses in the oceans as there are bacteria and archaea. Viruses are the main agents responsible for the rapid destruction of harmful algal blooms, which often kill other marine life. The number of viruses in the oceans decreases further offshore and deeper into the water, where there are fewer host organisms.

Microscopic organisms live in every part of the biosphere. The mass of prokaryote microorganisms — which includes bacteria and archaea, but not the nucleated eukaryote microorganisms — may be as much as 0.8 trillion tons of carbon (of the total biosphere mass, estimated at between 1 and 4 trillion tons). Barophilic marine microbes have been found at more than a depth of 10,000 m (33,000 ft; 6.2 mi) in the Mariana Trench, the deepest spot in the Earth's oceans. In fact, single-celled life forms have been found in the deepest part of the Mariana Trench, by the Challenger Deep, at depths of 11,034 m (36,201 ft; 6.856 mi). Other researchers reported related studies that microorganisms thrive inside rocks up to 580 m (1,900 ft; 0.36 mi) below the sea floor under 2,590 m (8,500 ft; 1.61 mi) of ocean off the coast of the northwestern United States, as well as 2,400 m (7,900 ft; 1.5 mi) beneath the seabed off Japan. The greatest known temperature at which microbial life can exist is 122 °C (252 °F) (*Methanopyrus kandleri*). On 20 August 2014, scientists confirmed the existence of microorganisms living 800 m (2,600 ft; 0.50 mi) below the ice of Antarctica. According to one researcher, "You can find microbes everywhere — they're extremely adaptable to conditions, and survive wherever they are."

Marine Viruses

A virus is a small infectious agent that replicates only inside the living cells of other organisms. Viruses can infect all types of life forms, from animals and plants to microorganisms, including bacteria and archaea.

When not inside an infected cell or in the process of infecting a cell, viruses exist in the form of independent particles. These viral particles, also known as *virions*, consist of two or three parts: (i) the genetic material made from either DNA or RNA, long molecules that carry genetic information; (ii) a protein coat, called the capsid, which surrounds and protects the genetic material; and in some cases (iii) an envelope of lipids that surrounds the protein coat when they are outside a cell. The shapes of these virus particles range from simple helical and icosahedral forms for some virus species to more complex structures for others. Most virus species have virions that are too small to be seen with an optical microscope. The average virion is about one one-hundredth the size of the average bacterium.

The origins of viruses in the evolutionary history of life are unclear: some may have evolved from plasmids—pieces of DNA that can move between cells—while others may have evolved from bacteria. In evolution, viruses are an important means of horizontal gene transfer, which increases genetic diversity. Viruses are considered by some to be a life form, because they carry genetic material, reproduce, and evolve through natural selection. However they lack key characteristics (such as cell structure) that are generally considered necessary to count as life. Because they possess some but not all such qualities, viruses have been described as "organisms at the edge of life" and as replicators.

Viruses are found wherever there is life and have probably existed since living cells first evolved. The origin of viruses is unclear because they do not form fossils, so molecular techniques have been used to compare the DNA or RNA of viruses and are a useful means of investigating how they arose.

Viruses are now recognised as ancient and as having origins that pre-date the divergence of life into the three domains.

Opinions differ on whether viruses are a form of life, or organic structures that interact with living organisms. They have been described as "organisms at the edge of life", since they resemble organisms in that they possess genes, evolve by natural selection, and reproduce by creating multiple copies of themselves through self-assembly. Although they have genes, they do not have a cellular structure, which is often seen as the basic unit of life. Viruses do not have their own metabolism, and require a host cell to make new products. They therefore cannot naturally reproduce outside a host cell.

Bacterial viruses, called bacteriophages, are a common and diverse group of viruses and are the most abundant form of biological entity in aquatic environments – there are up to ten times more of these viruses in the oceans than there are bacteria, reaching levels of 250,000,000 bacteriophages per millilitre of seawater.

There are also archaean viruses which replicate within archaea: these are double-stranded DNA viruses with unusual and sometimes unique shapes. These viruses have been studied in most detail in the thermophilic archaea, particularly the orders Sulfolobales and Thermoproteales.

A teaspoon of seawater contains about one million viruses. Most of these are bacteriophages, which are harmless to plants and animals, and are in fact essential to the regulation of saltwater and freshwater ecosystems. They infect and destroy bacteria in aquatic microbial communities, and are the most important mechanism of recycling carbon in the marine environment. The organic molecules released from the dead bacterial cells stimulate fresh bacterial and algal growth. Viral activity may also contribute to the biological pump, the process whereby carbon is sequestered in the deep ocean.

Microorganisms constitute more than 90% of the biomass in the sea. It is estimated that viruses kill approximately 20% of this biomass each day and that there are 15 times as many viruses in the oceans as there are bacteria and archaea. Viruses are the main agents responsible for the rapid destruction of harmful algal blooms, which often kill other marine life. The number of viruses in the oceans decreases further offshore and deeper into the water, where there are fewer host organisms.

Viruses are an important natural means of transferring genes between different species, which increases genetic diversity and drives evolution. It is thought that viruses played a central role in the early evolution, before the diversification of bacteria, archaea and eukaryotes, at the time of the last universal common ancestor of life on Earth. Viruses are still one of the largest reservoirs of unexplored genetic diversity on Earth.

Prokaryotes

Marine Bacteria

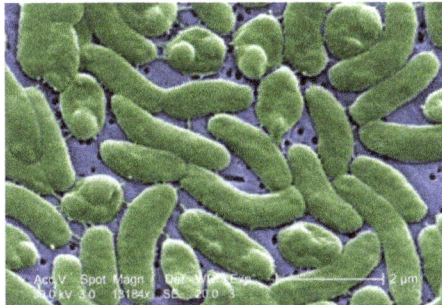

Vibrio vulnificus, a virulent bacterium found in estuaries and along coastal areas

Bacteria constitute a large domain of prokaryotic microorganisms. Typically a few micrometres in length, bacteria have a number of shapes, ranging from spheres to rods and spirals. Bacteria were among the first life forms to appear on Earth, and are present in most of its habitats. Bacteria inhabit soil, water, acidic hot springs, radioactive waste, and the deep portions of Earth's crust. Bacteria also live in symbiotic and parasitic relationships with plants and animals.

Once regarded as plants constituting the class *Schizomycetes*, bacteria are now classified as prokaryotes. Unlike cells of animals and other eukaryotes, bacterial cells do not contain a nucleus and rarely harbour membrane-bound organelles. Although the term *bacteria* traditionally included all prokaryotes, the scientific classification changed after the discovery in the 1990s that prokaryotes consist of two very different groups of organisms that evolved from an ancient common ancestor. These evolutionary domains are called *Bacteria* and *Archaea*.

The ancestors of modern bacteria were unicellular microorganisms that were the first forms of life to appear on Earth, about 4 billion years ago. For about 3 billion years, most organisms were microscopic, and bacteria and archaea were the dominant forms of life. Although bacterial fossils exist, such as stromatolites, their lack of distinctive morphology prevents them from being used to examine the history of bacterial evolution, or to date the time of origin of a particular bacterial species. However, gene sequences can be used to reconstruct the bacterial phylogeny, and these studies indicate that bacteria diverged first from the archaeal/eukaryotic lineage. Bacteria were also involved in the second great evolutionary divergence, that of the archaea and eukaryotes. Here, eukaryotes resulted from the entering of ancient bacteria into endosymbiotic associations with the ancestors of eukaryotic cells, which were themselves possibly related to the Archaea. This involved the engulfment by proto-eukaryotic cells of alphaproteobacterial symbionts to form either mitochondria or hydrogenosomes, which are still found in all known Eukarya. Later on, some eukaryotes that already contained mitochondria also engulfed cyanobacterial-like organisms. This led to the formation of chloroplasts in algae and plants. There are also some algae that originated from even later endosymbiotic events. Here, eukaryotes engulfed a eukaryotic algae that developed into a "second-generation" plastid. This is known as secondary endosymbiosis.

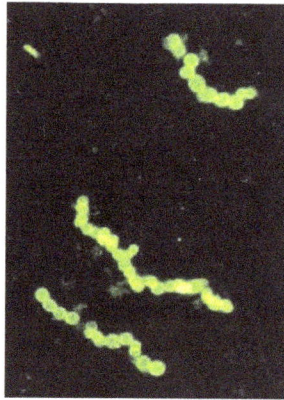

The marine *Thiomargarita namibiensis*, largest known bacterium

Cyanobacteria blooms can contain lethal cyanotoxins

Chloroplasts, such as the chloroplasts of this glaucophyte, may have originated from cyanobacteria.

Bacteria can be beneficial. This Pompeii worm, an extremophile found only at hydrothermal vents, has a protective cover of bacteria.

The largest known bacterium, the marine *Thiomargarita namibiensis*, can be visible to the naked eye and sometimes attains 0.75 mm (750 μm).

Marine Archaea

Archaea were initially viewed as extremophiles living in harsh environments, such as the yellow archaea pictured here in a hot spring, but they have since been found in a much broader range of habitats.

The archaea (Greek for *ancient*) constitute a domain and kingdom of single-celled microorganisms. These microbes are prokaryotes, meaning they have no cell nucleus or any other membrane-bound organelles in their cells.

Archaea were initially classified as bacteria, but this classification is outdated. Archaeal cells have unique properties separating them from the other two domains of life, Bacteria and Eukaryota. The Archaea are further divided into multiple recognized phyla. Classification is difficult because the majority have not been isolated in the laboratory and have only been detected by analysis of their nucleic acids in samples from their environment.

Archaea and bacteria are generally similar in size and shape, although a few archaea have very strange shapes, such as the flat and square-shaped cells of *Haloquadratum walsbyi*. Despite this morphological similarity to bacteria, archaea possess genes and several metabolic pathways that are more closely related to those of eukaryotes, notably the enzymes involved in transcription and translation. Other aspects of archaeal biochemistry are unique, such as their reliance on ether lipids in their cell membranes, such as archaeols. Archaea use more energy sources than eukaryotes: these range from organic compounds, such as sugars, to ammonia, metal ions or even hydrogen gas. Salt-tolerant archaea (the Haloarchaea) use sunlight as an energy source, and other species of archaea fix carbon; however, unlike plants and cyanobacteria, no known species of archaea does both. Archaea reproduce asexually by binary fission, fragmentation, or budding; unlike bacteria and eukaryotes, no known species forms spores. Unlike viruses and bacteria, no known archaea is a pathogen.

Archaea are particularly numerous in the oceans, and the archaea in plankton may be one of the most abundant groups of organisms on the planet. Archaea are a major part of Earth's life and may play roles in both the carbon cycle and the nitrogen cycle.

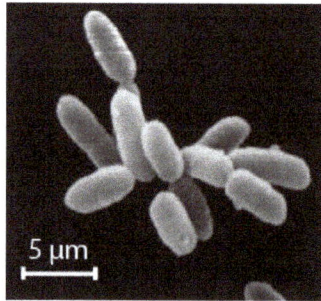

Halobacteria, found in water saturated or nearly saturated with salt, are now recognized as being archaea.

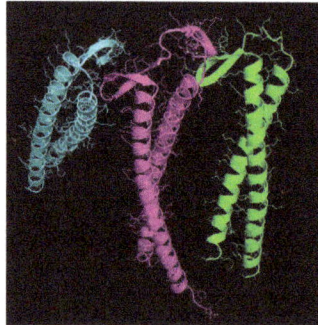

Crystal structure of prefoldin from the heat-loving marine archaea *Pyrococcus horikoshii*

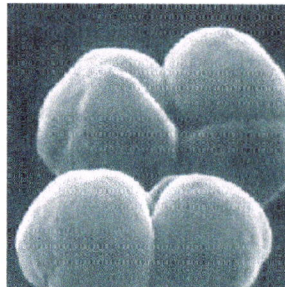

Methanosarcina barkeri, a marine archaea that produces methane

Eukaryotes

Marine Fungi

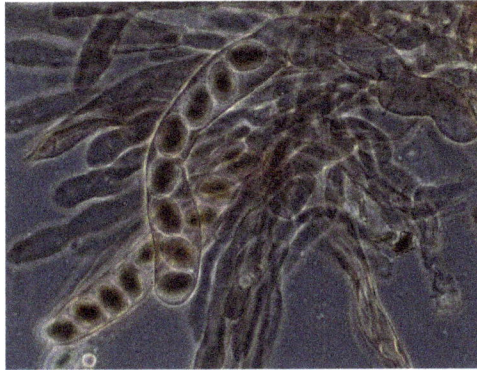

Marine Ascomycete fungus

Lichen on a rock in a marine splash zone. Lichens are mutualistic associations between a fungus and an alga or cyanobacterium.

Over 1500 species of fungi are known from marine environments. These are parasitic on marine algae or animals, or are saprobes on algae, corals, protozoan cysts, sea grasses, wood and other substrata, and can also be found in sea foam. Spores of many species have special appendages which facilitate attachment to the substratum. A very diverse range of unusual secondary metabolites is produced by marine fungi.

Mycoplankton are saprotropic members of the plankton communities of marine and freshwater ecosystems. They are composed of filamentous free-living fungi and yeasts that are associated with planktonic particles or phytoplankton. Similar to bacterioplankton, these aquatic fungi play a significant role in heterotrophic mineralization and nutrient cycling. Mycoplankton can be up to 20 mm in diameter and over 50 mm in length.

In a typical milliliter of seawater, there are approximately 10^3 to 10^4 fungal cells. This number is greater in coastal ecosystems and estuaries due to nutritional runoff from terrestrial communities. The greatest diversity and number of species of mycoplankton is found in surface waters (< 1000 m), and the vertical profile depends on the abundance of phytoplankton. Furthermore, this difference in distribution may vary between seasons due to nutrient availability. Marine fungi survive in a constant oxygen deficient environment, and therefore depend on oxygen diffusion by turbulence and oxygen generated by photosynthetic organisms.

Marine fungi can be classified as:

- Lower fungi - adapted to marine habitats (zoosporic fungi, including mastigomycetes: oomycetes and chytridiomycetes)

- Higher fungi - filamentous, modified to planktonic lifestyle (hyphomycetes, ascomycetes, basidiomycetes)

Most mycoplankton species are higher fungi.

Lichens are mutualistic associations between a fungus, usually an ascomycete, and an alga or a cyanobacterium. Several lichens are found in marine environments. Many more occur in the splash zone, where they occupy different vertical zones depending on how tolerant they are to submersion. Fossil marine lichens 600 million years old have been discovered in the late Neoproterozoic marine phosphate rocks in the sedimentary, fossil-rich Doushantuo Formation in China.

According to fossil records, fungi date back to the late Proterozoic era 900-570 million years ago. It has been hypothesized that mycoplankton evolved from terrestrial fungi, likely in the Paleozoic era (390 million years ago).

Marine Protists

Of eukaryotic groups, the protists are most commonly unicellular and microscopic. This is a highly diverse group of organisms that are not easy to classify. Several algae species are multicellular protists, and there are marine slime molds have unique life cycles that involve switching between unicellular, colonial, and multicellular forms. The number of species of protists is unknown, and we may have identified only a small portion. Studies from 2001-2004 have shown that a high degree of protist diversity exists in oceans, deep sea-vents, river sediment and an acidic river which suggests that a large number of eukaryotic microbial communities have yet to be discovered.

Marine Algae

Algae can grow as single cells, or in long chains of cells. Green algae are a large group of photosynthetic eukaryotes that include many microscopic organisms. Green algae includes unicellular and colonial flagellates as well as various colonial, coccoid, and filamentous forms. There are about 6000 species.

Marine Animals

At least one microscopic animal group, the parasitic cnidarian Myxozoa, is unicellular in its adult form, and includes marine species. Other marine micro-animals are multicellular. Microscopic arthropods are more commonly found inland in freshwater, but there are marine species as well. Microscopic marine crustaceans include some copepods, cladocera and water bears. Some marine nematodes and rotifers are also too small to be seen with the naked eye, as are many loricifera, including the recently discovered anaerobic species which spend their lives in an anoxic environment. Copepods contribute more to the secondary productivity and carbon sink of the world oceans than any other group of organisms.

- Marine micro-animals

Armoured *Pliciloricus enigmaticus* of the phylum Loricifera live in the spaces between marine gravel

Marine Mammal

A humpback whale (Megaptera novaeangliae), a member of infraorder Cetacea of the order Cetartiodactyla

Marine mammals are aquatic mammals that rely on the ocean and other marine ecosystems for their existence. They include animals such as seals, whales, manatees, sea otters and polar bears. They do not represent a distinct taxon or systematic grouping, but rather have a paraphyletic relation. They are also unified by their reliance on the marine environment for feeding.

A leopard seal (*Hydrurga leptonyx*), a member of the clade Pinnipedia of the order Carnivora

Marine mammal adaptation to an aquatic lifestyle vary considerably between species. Both cetaceans and sirenians are fully aquatic and therefore are obligate water dwellers. Seals and sea-lions are semiaquatic; they spend the majority of their time in the water, but need to return to land for important activities such as mating, breeding and molting. In contrast, both otters and the polar bear are much less adapted to aquatic living. Their diet varies considerably as well; some may eat zooplankton, others may eat fish, squid, shellfish, sea-grass and a few may eat other mammals. While the number of marine mammals is small compared to those found on land, their roles in various ecosystems are large, especially concerning the maintanence of marine ecosystems, through processes including the regulation of prey populations. This role in maintaining ecosystems makes them of particular concern as 23% of marine mammal species are currently threatened.

Marine mammals were first hunted by aboriginal peoples for food and other resources. Many were also the target for commercial industry, leading to a sharp decline in all populations of exploited species, such as whales and seals. Commercial hunting lead to the extinction of †Steller's sea cow and the †Caribbean monk seal. After commercial hunting ended, some species, such as the gray whale and northern elephant seal, have rebounded in numbers; conversely, other species, such as the North Atlantic right whale, are critically endangered. Other than hunting, marine mammals can be killed as bycatch from fisheries, where they become entangled in fixed netting and drown or starve. Increased ocean traffic causes collisions between fast ocean vessels and large marine mammals. Habitat degradation also threatens marine mammals and their ability to find and catch food. Noise pollution, for example, may adversely affect echolocating mammals, and the ongoing effects of global warming degrade arctic environments.

Taxonomy

Marine mammals vary greatly in size and shape

Marine mammals form a diverse group of 129 species that rely on the ocean for their existence. They do not represent a distinct taxon or systematic grouping, but instead have a paraphyletic relationship. They are also unified by their reliance on the marine environment for feeding. The level of dependence on the marine environment for existence varies considerably with species. For example, dolphins and whales are completely dependent on the marine environment for all stages of their life, seals feed in the ocean but breed on land, and polar bears must feed on land. Twenty three percent of marine mammal species are threatened.

A polar bear (*Ursus maritimus*), a member of family Ursidae.

A sea otter (*Enhydra lutris*), a member of family Mustelidae.

California sea lions (*Zalophus californianus*), members of the family Otariidae.

A West Indian manatee (*Trichechus manatus*), a member of order Sirenia.

A southern right whale (*Eubalaena australis*), a member of the order Cetartiodactyla.

Classification of Extant Species

- Order Cetartiodactyla

  o Suborder Whippomorpha

    - Family Balaenidae (right and bowhead whales), two genera and four species

    - Family Cetotheriidae (pygmy right whale), one species

    - Family Balaenopteridae (rorquals), two genera and eight species

    - Family Eschrichtiidae (gray whale), one species

    - Family Physeteridae (sperm whale), one species

    - Family Kogiidae (pygmy and dwarf sperm whales), one genus and two species

- Family Monodontidae (narwhal and beluga), two genera and two species

- Family Ziphiidae (beaked whales), six genera and 21 species

- Family Delphinidae (oceanic dolphins), 17 genera and 38 species

- Family Phocoenidae (porpoises), two genera and seven species

- Order Sirenia (sea cows)

    o   Suborder Cynodontia

    - Family Trichechidae (manatees), one species

    - Family Dugongidae (dugongs), one species

- Order Carnivora (carnivores):

    o   Suborder Caniformia

    - Family Mustelidae, two species

    - Family Ursidae (bears), one species

    o   Suborder Pinnipedia (sealions, walruses, seals)

    - Family Otariidae (eared seals), seven genera and 15 species

    - Family Odobenidae (walrus), one species

    - Family Phocidae (earless seals), 14 genera and 18 species

Evolution

A skeleton of †*Thalassocnus* (5–3 mya) from the Muséum national d'histoire naturelle in its presumed swimming pose

Mammals have returned to the ocean in many separate evolutionary lineages, namely: Cetacea, Sirenia, †Desmostylia, Pinnipedia, †*Kolponomos* (marine bear),

†*Thalassocnus* (aquatic sloth), *Ursus maritimus* (polar bear), and *Enhydra lutris* (sea otter); the eutriconodonts †*Ichthyoconodon* and †*Dyskritodon* might have also been marine in habits. Five of these lineages are extinct (†Desmostylia; †*Kolponomos*; †*Thalassocnus*, †*Dyskritodon*, †*Ichthyoconodon*). Despite the diversity in morphology seen between groups, improving foraging efficiency has been the main driver in the evolution in these lineages. Today, fully aquatic marine mammals belong to one of two orders: Cetartiodactyla or Sirenia. The Cetartiodactyla lineage became aquatic around 50 million years ago (mya), then Sirenia 40 mya, and Pinnipedia around 20 to 25 mya, then sea otters two mya, and most recently the polar bear around 130,000 to 110,000 years ago.

Based on molecular and morphological research, the cetaceans genetically and morphologically fall firmly within the Artiodactyla (even-toed ungulates). The term Cetartiodactyla reflects the idea that whales evolved within the ungulates. The term was coined by merging the name for the two orders, Cetacea and Artiodactyla, into a single word. Under this definition, the closest living land relative of the whales and dolphins is thought to be the hippopotamuses. Use of the order Cetartiodactyla, instead of Cetacea with parvorders Odontoceti and Mysticeti, is favored by most evolutionary mammalogists working with molecular data and is supported the IUCN Cetacean Specialist Group and by Taxonomy Committee of the Society for Marine Mammalogy, the largest international association of marine mammal scientists in the world. Some others, including many marine mammalogists and paleontologists, favor retention of order Cetacea with the two suborders in the interest of taxonomic stability.

Pinnipeds split from other caniforms 50 mya during the Eocene. Their evolutionary link to terrestrial mammals was unknown until the 2007 discovery of †*Puijila darwini* in early Miocene deposits in Nunavut, Canada. Like a modern otter, †*Puijila* had a long tail, short limbs and webbed feet instead of flippers. The lineages of Otariidae (eared seals) and Odobenidae (walrus) split almost 28 mya. Phocids (earless seals) are known to have existed for at least 15 mya, and molecular evidence supports a divergence of the Monachinae (monk seals) and Phocinae lineages 22 mya.

Fossil evidence indicates the sea otter (*Enhydra*) lineage became isolated in the North Pacific approximately two mya, giving rise to the now-extinct †*Enhydra macrodonta* and the modern sea otter, *Enhydra lutris*. The sea otter evolved initially in northern Hokkaidō and Russia, and then spread east to the Aleutian Islands, mainland Alaska, and down the North American coast. In comparison to cetaceans, sirenians, and pinnipeds, which entered the water approximately 50, 40, and 20 mya, respectively, the sea otter is a relative newcomer to marine life. In some respects though, the sea otter is more fully adapted to water than pinnipeds, which must haul out on land or ice to give birth.

Illustration of †*Prorastomus*, the most known primitive sirenian (40 mya)

The first appearance of sirenians in the fossil record was during the early Eocene, and by the late Eocene, sirenians had significantly diversified. Inhabitants of rivers, estuaries, and nearshore marine waters, they were able to spread rapidly. The most primitive sirenian, †*Prorastomus*, was found in Jamaica, unlike other marine mammals which originated from the Old World (such as cetaceans). The first known quadrupedal sirenian was †*Pezosiren* from the early Eocene. The earliest known sea cows, of the families †Prorastomidae and †Protosirenidae, were both confined to the Eocene, and were pig-sized, four-legged, amphibious creatures. The first members of Dugongidae appeared by the end of the Eocene. At this point, sea cows were fully aquatic.

Polar bears are thought to have diverged from a population of brown bears, *Ursus arctos*, that became isolated during a period of glaciation in the Pleistocene or from the eastern part of Siberia, (from Kamchatka and the Kolym Peninsula). The oldest known polar bear fossil is a 130,000 to 110,000-year-old jaw bone, found on Prince Charles Foreland in 2004. The mitochondrial DNA (mtDNA) of the polar bear diverged from the brown bear roughly 150,000 years ago. Further, some clades of brown bear, as assessed by their mtDNA, are more closely related to polar bears than to other brown bears, meaning that the polar bear might not be considered a species under some species concepts.

Adaptations

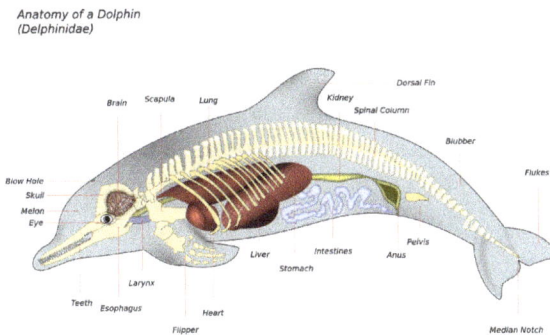

The anatomy of a dolphin showing its skeleton, major organs, and body shape

Marine mammals have a number of physiological and anatomical features to overcome the unique challenges associated with aquatic living. Some of these features are very species specific. Marine mammals have developed a number of features for efficient locomotion such as torpedo shaped bodies to reduce drag; modified limbs for propulsion and steering; tail flukes and dorsal fins for propulsion and balance. Marine mammals are adept at thermoregulation using dense fur or blubber, circulatory adjustments (counter-current heat exchangers); torpedo shaped bodies, reduced appendages, and large size to prevent heat loss.

Marine mammals are able to dive for long periods of time. Both pinnipeds and cetaceans have large and complex blood vessel systems which serve to store oxygen to support deep diving. Other important reservoirs include muscles, blood, and the spleen which all have the capacity to hold a high concentration of oxygen. They are also capable of bradycardia (reduced heart rate), and vasoconstriction (shunting most of the oxygen to vital organs such as the brain and heart) to allow extended diving times and cope with oxygen deprivation. If oxygen is depleted, marine mammals can access substantial reservoirs of glycogen that support anaerobic glycolysis of the cells involved during conditions of systemic hypoxia associated with prolonged submersion.

Sound travels differently through water therefore marine mammals have developed adaptations to ensure effective communication, prey capture, and predator detection. The most notable adaptation is the development of echolocation in whales and dolphins.

Marine mammals have evolved a wide variety of features for feeding, which are mainly seen in their dentition. For example, the cheek teeth of pinniped and odontocetes are specifically adapted to capture fish and squid. In contrast, baleen whales have evolved baleen plates to filter feed plankton and small fish from the water.

Polar bears, otters, and fur seals have fur, one of the defining mammalian features, that is long, oily, and waterproof in order to trap air to provide insulation. In contrast, other marine mammals – such as whales, dolphins, porpoises, manatees, dugongs, and walruses – have lost long fur in favor of a thick, dense epidermis and a thickened fat layer (blubber) in response to hydrodynamic requirements. Wading and bottom-feeding animals (such as manatee) need to be heavier than water in order to keep contact with the floor or to stay submerged. Surface-living animals (such as sea otters) need the opposite, and free-swimming animals living in open waters (such as dolphins) need to be neutrally buoyant in order to be able to swim up and down the water column. Typically, thick and dense bone is found in bottom feeders and low bone density is associated with mammals living in deep water. Some marine mammals, such as polar bears and otters, have retained four weight-bearing limbs and can walk on land like fully terrestrial animals.

Distribution and Habitat

Marine mammals are widely distributed throughout the globe, but their distribution is patchy and coincides with the productivity of the oceans. Species richness peaks at around 40° latitude, both north and south. This corresponds to the highest levels of primary production around North and South America, Africa, Asia and Australia. Total species range is highly variable for marine mammal species. On average most marine mammals have ranges which are equivalent or smaller than one-fifth of the Indian Ocean. The variation observed in range size is a result of the different ecological requirements of each species and their ability to cope with a broad range of environmental conditions. The high degree of overlap between marine mammal species richness and areas of human impact on the environment is of concern.

Bottlenose dolphin at Dolphin Reef, Eilat, Israel

Most marine mammals, such as seals and sea otters, inhabit the coast. Seals, however, also use a number of terrestrial habitats, both continental and island. In temperate and tropical areas, they haul-out on to sandy and pebble beaches, rocky shores, shoals, mud flats, tide pools and in sea caves. Some species also rest on man-made structures, like piers, jetties, buoys and oil platforms. Seals may move further inland and rest in sand dunes or vegetation, and may even climb cliffs. Most cetaceans live in the open ocean, and species like the sperm whale may dive to depths of –1,000 to –2,500 feet (–300 to –760 m) in search of food. Sirenians live in shallow coastal waters, usually living 30 feet (9.1 m) below sea level. However, they have been known to dive to –120 feet (–37 m) to forage deep-water seagrasses. Sea otters live in protected areas, such as rocky shores, kelp forests, and barrier reefs, although they may reside among drift ice or in sandy, muddy, or silty areas.

Many marine mammals seasonally migrate. Annual ice contains areas of water that appear and disappear throughout the year as the weather changes, and seals migrate in response to these changes. In turn, polar bears must follow their prey. In Hudson Bay, James Bay, and some other areas, the ice melts completely each summer (an event often referred to as "ice-floe breakup"), forcing polar bears to go onto land and wait through the months until the next freeze-up. In the Chukchi and Beaufort seas, polar bears retreat each summer to the ice further north that remains frozen year-round. Seals may also migrate to other environmental changes, such as El Niño, and traveling seals may use various features of their environment to reach their destination including geomagnetic fields, water and wind currents, the position of the sun and moon and the taste and temperature of the water. Baleen whales famously migrate very long distances into tropical waters to give birth and raise young, possibly to prevent predation by killer whales. The gray whale has the longest recorded migration of any mammal, with one traveling 14,000 miles (23,000 km) from the Sea of Okhotsk to the Baja Peninsula. During the winter, manatees living at the northern end of their range migrate to warmer waters.

Ecology

Diet

All cetaceans are carnivorous and predatory. Toothed whales mostly feed on fish and cephalopods, followed by crustaceans and bivalves. Some may forage with other kinds of animals, such as other species of whales or certain species of pinnipeds. One common feeding method is herding, where a pod squeezes a school of fish into a small volume, known as a bait ball. Individual members then take turns plowing through the ball, feeding on the stunned fish. Coralling is a method where dolphins chase fish into shallow water to catch them more easily. Killer whales and bottlenose dolphins have also been known to drive their prey onto a beach to feed on it. Other whales with a blunt snout and reduced dentition rely on suction feeding. Though carnivorous, they house gut flora similar to that of terrestrial herbivores, probably a remnant of their herbivorous ancestry.

Killer whale hunting a Weddel seal

Baleen whales use their baleen plates to sieve plankton, among others, out of the water; there are two types of methods: lunge-feeding and gulp-feeding. Lunge-feeders expand the volume of their jaw to a volume bigger than the original volume of the whale itself by inflating their mouth. This causes grooves on their throat to expand, increasing the amount of water the mouth can store. They ram a baitball at high speeds in order to feed, but this is only energy-effective when used against a large baitball. Gulp-feeders swim with an open mouth, filling it with water and prey. Prey must occur in sufficient numbers to trigger the whale's interest, be within a certain size range so that the baleen plates can filter it, and be slow enough so that it cannot escape.

Sea otters have dexterous hands which they use to smash sea urchins off of rocks.

Otters are the only marine animals that are capable of lifting and turning over rocks, which they often do with their front paws when searching for prey. The sea otter may pluck snails and other organisms from kelp and dig deep into underwater mud for clams. It is the only marine mammal that catches fish with its forepaws rather than with its teeth. Under each foreleg, sea otters have a loose pouch of skin that extends across the chest which they use to store collected food to bring to the surface. This pouch also holds a rock that is used to break open shellfish and clams, an example of tool use. The sea otters eat while floating on their backs, using their forepaws to tear food apart and bring to their mouths. Marine otters mainly feed on crustaceans and fish.

Pinnipeds mostly feed on fish and cephalopods, followed by crustaceans and bivalves, and then zooplankton and warm-blooded prey (like sea birds). Most species are generalist feeders, but a few are specialists. They typically hunt non-schooling fish, slow-moving or immobile invertebrates or endothermic prey when in groups. Solitary foraging species usually exploit coastal waters, bays and rivers. When large schools of fish or squid are available, pinnipeds hunt cooperatively in large groups, locating and herding their prey. Some species, such as California and South American sea lions, may forage with cetaceans and sea birds.

The polar bear is the most carnivorous species of bear, and its diet primarily consists of ringed (*Pusa hispida*) and bearded (*Erignathus barbatus*) seals. Polar bears hunt primarily at the interface between ice, water, and air; they only rarely catch seals on land or in open water. The polar bear's most common hunting method is still-hunting: The bear locates a seal breathing hole using its sense of smell, and crouches nearby for a seal to appear. When the seal exhales, the bear smells its breath, reaches into the hole with a forepaw, and drags it out onto the ice. The polar bear also hunts by stalking seals resting on the ice. Upon spotting a seal, it walks to within 100 yards (90 m), and then crouches. If the seal does not notice, the bear creeps to within 30 to 40 feet (9 to 10 m) of the seal and then suddenly rushes to attack. A third hunting method is to raid the birth lairs that female seals create in the snow. They may also feed on fish.

A dugong feeding on the sea-floor

Sirenians are referred to as "sea cows" because their diet consists mainly of sea-grass. When eating, they ingest the whole plant, including the roots, although when this is impossible they feed on just the leaves. A wide variety of seagrass has been found in dugong stomach contents, and evidence exists they will eat algae when seagrass is scarce. West Indian manatees eat up to 60 different species of plants, as well as fish and small invertebrates to a lesser extent.

Keystone Species

Sea otters are a classic example of a keystone species; their presence affects the ecosystem more profoundly than their size and numbers would suggest. They keep the population of certain benthic (sea floor) herbivores, particularly sea urchins, in check. Sea urchins graze on the lower stems of kelp, causing the kelp to drift away and die. Loss of the habitat and nutrients provided by kelp forests leads to profound cascade effects on the marine ecosystem. North Pacific areas that do not have sea otters often turn into urchin barrens, with abundant sea urchins and no kelp forest. Reintroduction of sea otters to British Columbia has led to a dramatic improvement in the health of coastal ecosystems, and similar changes have been observed as sea otter populations recovered in the Aleutian and Commander Islands and the Big Sur coast of California However, some kelp forest ecosystems in California have also thrived without sea otters, with sea urchin populations apparently controlled by other factors. The role of sea otters in maintaining kelp forests has been observed to be more important in areas of open coast than in more protected bays and estuaries.

Antarctic fur seal pups (left) vs. Arctic harp seal pup (right)

An apex predator affects prey population dynamics and defense tactics (such as camouflage). The polar bear is the apex predator within its range. Several animal species, particularly Arctic foxes (*Vulpes lagopus*) and glaucous gulls (*Larus hyperboreus*), routinely scavenge polar bear kills. The relationship between ringed seals and polar bears is so close that the abundance of ringed seals in some areas appears to regulate the density of polar bears, while polar bear predation in turn regulates density and reproductive success of ringed seals. The evolutionary pressure of polar bear predation on seals probably accounts for some significant differences between Arctic and Antarctic seals. Compared to the Antarctic, where there is no major surface predator, Arctic seals use more breathing holes per individual, appear more restless when hauled out on the ice, and rarely defecate on the ice. The fur of Arctic pups is white, presumably to provide camouflage from predators, whereas Antarctic pups all have dark fur.

Killer whales are apex predators throughout their global distribution, and can have a profound effect on the behavior and population of prey species. Their diet is very broad and they can feed on many vertebrates in the ocean including salmon, rays, sharks (even white sharks), large baleen whales, and nearly 20 species of pinniped. The predation of whale calves may be responsible for annual whale migrations to calving grounds in more tropical waters, where the population of killer whales is much lower than in polar waters. Prior to whaling, it is thought that great whales were a major food source; however, after their sharp decline, killer whales have since expanded their diet, leading to the decline of smaller marine mammals. A decline in Aleutian Islands sea otter populations in the 1990s was controversially attributed by some scientists to killer whale predation, although with no direct evidence. The decline of sea otters followed a decline in harbor seal and Steller sea lion populations, the killer whale's preferred prey, which in turn may be substitutes for their original prey, now reduced by industrial whaling.

Whale Pump

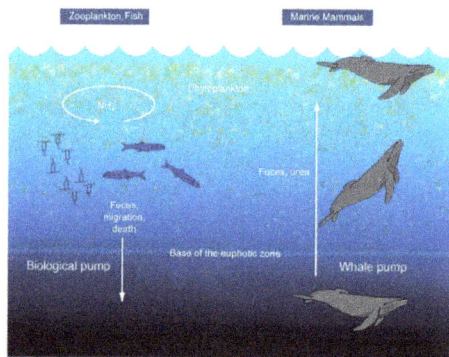

"Whale pump" – the role played by whales in recycling ocean nutrients

A 2010 study considered whales to be a positive influence to the productivity of ocean fisheries, in what has been termed a "whale pump". Whales carry nutrients such as nitrogen from the depths back to the surface. This functions as an upward biological pump, reversing an earlier presumption that whales accelerate the loss of nutrients to the bottom. This nitrogen input in the Gulf of Maine is more than the input of all rivers combined emptying into the gulf, some 25,000 short tons (23,000 t) each year. Whales defecate at the ocean's surface; their excrement is important for fisheries because it is rich in iron and nitrogen. The whale feces are liquid and instead of sinking, they stay at the surface where phytoplankton feed off it.

Upon death, whale carcasses fall to the deep ocean and provide a substantial habitat for marine life. Evidence of whale falls in present-day and fossil records shows that deep sea whale falls support a rich assemblage of creatures, with a global diversity of 407 species, comparable to other neritic biodiversity hotspots, such as cold seeps and hydrothermal vents. Deterioration of whale carcasses happens though a series of three stages. Initially, moving organisms, such as sharks and hagfish, scavenge soft tissue at a rapid rate over a period of months to as long as two years. This is followed by the colonization of bones and surrounding sediments (which contain organic matter) by enrichment opportunists, such as crustaceans and polychaetes, throughout a period of years. Finally, sulfophilic bacteria reduce the bones releasing hydrogen sulphide enabling the growth of chemoautotrophic organisms, which in turn, support other organisms such as mussels, clams, limpets, and sea snails. This stage may last for decades and supports a rich assemblage of species, averaging 185 species per site.

Interactions with Humans

Exploitation

Men killing northern fur seals on Saint Paul Island, Alaska in the 1890s

Marine mammals were hunted by coastal aboriginal humans historically for food and other resources. These subsistence hunts still occur in Canada, Greenland, Indonesia, Russia, the United States, and several nations in the Caribbean. The effects of these are only localized, as hunting efforts were on a relatively small scale. Commercial hunting took this to a much greater scale and marine mammals were heavily exploited. This led to the extinction of the †Steller's sea cow (along with subsistence hunting) and the †Caribbean monk seal. Today, populations of species that were historically hunted, such as blue whales (*Balaenoptera musculus*) and the North Pacific right whale (*Eubalaena japonica*), are much lower than their pre-whaling levels. Because whales generally have slow growth rates, are slow to reach sexual maturity, and have a low reproductive output, population recovery has been very slow.

A number of whales are still subject to direct hunting, despite the 1986 moratorium ban on whaling set under the terms of the International Whaling Commission (IWC). There are only two nations remaining which sanction commercial whaling: Norway, where several hundred common minke whales are harvested each year; and Iceland, where quotas of 150 fin whales and 100 minke whales per year are set. Japan also harvests several hundred Antarctic and North Pacific minke whales each year, ostensibly for scientific research in accordance with the moratorium. However, the illegal trade of whale and dolphin meat is a significant market in Japan and some countries.

The most profitable furs in the fur trade were those of sea otters, especially the northern sea otter which inhabited the coastal waters between the Columbia River to the south and Cook Inlet to the north. The fur of the Californian southern sea otter was less highly prized and thus less profitable. After the northern sea otter was hunted to local extinction, maritime fur traders shifted to California until the southern sea otter was likewise nearly extinct. The British and American maritime fur traders took their furs to the Chinese port of Guangzhou (Canton), where they worked within the established

Canton System. Furs from Russian America were mostly sold to China via the Mongolian trading town of Kyakhta, which had been opened to Russian trade by the 1727 Treaty of Kyakhta.

Historical and modern range of northern sea otters

Commercial sealing was historically just as important as the whaling industry. Exploited species included harp seals, hooded seals, Caspian seals, elephant seals, walruses and all species of fur seal. The scale of seal harvesting decreased substantially after the 1960s, after the Canadian government reduced the length of the hunting season and implemented measures to protect adult females. Several species that were commercially exploited have rebounded in numbers; for example, Antarctic fur seals may be as numerous as they were prior to harvesting. The northern elephant seal was hunted to near extinction in the late 19th century, with only a small population remaining on Guadalupe Island. It has since recolonized much of its historic range, but has a population bottleneck. Conversely, the Mediterranean monk seal was extirpated from much of its former range, which stretched from the Mediterranean to the Black Sea and northwest Africa, and only remains in the northeastern Mediterranean and some parts of northwest Africa.

Ocean Traffic and Fisheries

The remains of a North Atlantic right whale after it collided with a ship propeller

By-catch is the incidental capture of non-target species in fisheries. Fixed and drift gill nets cause the highest mortality levels for both cetaceans and pinnipeds, however, entanglements in long lines, mid-water trawls, and both trap and pot lines are also common. Tuna seines are particularly problematic for entanglement by dolphins. By-catch affects all cetaceans, both small and big, in all habitat types. However, smaller cetaceans and pinnipeds are most vulnerable as their size means that escape once they are entangled is highly unlikely and they frequently drown. While larger cetaceans are capable of dragging nets with them, the nets sometimes remain tightly attached to the individual and can impede the animal from feeding sometimes leading to starvation. Abandoned or lost nets and lines cause mortality through ingestion or entanglement. Marine mammals also get entangled in aquaculture nets, however, these are rare events and not prevalent enough to impact populations.

Vessel strikes cause death for a number of marine mammals, especially whales. In particular, fast commercial vessels such as container ships can cause major injuries or death when they collide with marine mammals. Collisions occur both with large commercial vessels and recreational boats and cause injury to whales or smaller cetaceans. The critically endangered North Atlantic right whale is particularly affected by vessel strikes. Tourism boats designed for whale and dolphin watching can also negatively impact on marine mammals by interfering with their natural behavior.

The fishery industry not only threatens marine mammals through by-catch, but also through competition for food. Large scale fisheries have led to the depletion of fish stocks that are important prey species for marine mammals. Pinnipeds have been especially affected by the direct loss of food supplies and in some cases the harvesting of fish has led to food shortages or dietary deficiencies, starvation of young, and reduced recruitment into the population. As the fish stocks have been depleted, the competition between marine mammals and fisheries has sometimes led to conflict. Large-scale culling of populations of marine mammals by commercial fishers has been initiated in a number of areas in order to protect fish stocks for human consumption.

Shellfish aquaculture takes up space so in effect creates competition for space. However, there is little direct competition for aquaculture shellfish harvest. On the other hand, marine mammals regularly take finfish from farms, which creates significant problems for marine farmers. While there are usually legal mechanisms designed to deter marine mammals, such as anti-predator nets or harassment devices, individuals are often illegally shot.

Habitat Loss and Degradation

Habitat degradation is caused by a number of human activities. Marine mammals that live in coastal environments are most likely to be affected by habitat degradation and loss. Developments such as sewage marine outfalls, moorings, dredging, blasting, dumping, port construction, hydroelectric projects, and aquaculture both degrade the environment and take up valuable habitat. For example, extensive shellfish aquaculture takes up valuable space used by coastal marine mammals for important activities such as breeding, foraging and resting.

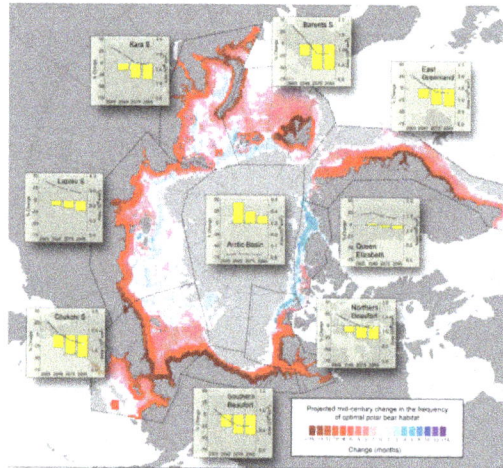

Map from the U.S. Geological Survey shows projected changes in polar bear habitat from 2005 to 2095. Red areas indicate loss of optimal polar bear habitat; blue areas indicate gain.

Contaminants that are discharged into the marine environment accumulate in the bodies of marine mammals when they are stored unintentionally in their blubber along with energy. Contaminants that are found in the tissues of marine mammals include heavy metals, such as mercury and lead, but also organochlorides and polycyclic aromatic hydrocarbons. For example, these can cause disruptive effects on endocrine systems; impair the reproductive system, and lower the immune system of individuals, leading to a higher number of deaths. Other pollutants such as oil, plastic debris and sewage threaten the livelihood of marine mammals.

Noise pollution from anthropogenic activities is another major concern for marine mammals. This is a problem because underwater noise pollution interferes with the abilities of some marine mammals to communicate, and locate both predators and prey. Underwater explosions are used for a variety of purposes including military activities, construction and oceanographic or geophysical research. They can cause injuries such as hemorrhaging of the lungs, and contusion and ulceration of the gastrointestinal tract. Underwater noise is generated from shipping, the oil and gas industry, research, and military use of sonar and oceanographic acoustic experimentation. Acoustic harassment devices and acoustic deterrent devices used by aquaculture facilities to scare away marine mammals emit loud and noxious underwater sounds.

Two changes to the global atmosphere due to anthropogenic activity threaten marine mammals. The first is increases in ultraviolet radiation due to ozone depletion, and this mainly affects the Antarctic and other areas of the southern hemisphere. An increase in ultraviolet radiation has the capacity to decrease phytoplankton abundance, which forms the basis of the food chain in the ocean. The second effect of global climate change is global warming due to increased carbon dioxide levels in the atmosphere. Raised sea levels, sea temperature and changed currents are expected to affect marine mammals by altering the distribution of important prey species, and changing the suitability of breeding sites and migratory routes. The Arctic food chain would be disrupted by the near extinction or migration of polar bears. Arctic sea ice is the polar bear's habitat. It has been declining at a rate of 13% per decade because the temperature is rising at twice the rate of the rest of the world. By the year 2050, up to two-thirds of the world's polar bears may vanish if the sea ice continues to melt at its current rate.

Protection

The Marine Mammal Protection Act of 1972 (MMPA) was passed on October 21, 1972 under president Richard Nixon to prevent the further depletion and possible extinction of marine mammal stocks. It prohibits the taking ("the act of hunting, killing, capture, and/or harassment of any marine mammal; or, the attempt at such") of any marine mammal without a permit issued by the Secretary. Authority to manage the MMPA was divided between the Secretary of the Interior through the U.S. Fish and Wildlife Service (Service), and the Secretary of Commerce, which is delegated to the National Oceanic and Atmospheric Administration (NOAA). The Marine Mammal Commission (MMC) was established to review existing policies and make recommendations to the Service and NOAA to better implement the MMPA. The Service is responsible for ensuring the protection of sea otters and marine otters, walruses, polar bears, the three species of manatees, and dugongs; and NOAA was given responsibility to conserve and manage pinnipeds (excluding walruses) and cetaceans.

The 1979 Convention on the Conservation of Migratory Species of Wild Animals (CMS) is the only global organization that conserves a broad range of animals, of which includes marine mammals. Of the agreements made, three of them deal with the conservation of marine mammals: ACCOBAMS, ASCOBANS, and the Wadden Sea Agreement.

The Agreement on the Conservation of Cetaceans in the Black Sea, Mediterranean Sea and contiguous Atlantic area (ACCOBAMS), founded in 1996, specifically protects cetaceans in the Mediterranean area, and "maintains a favorable status". There are 23 member states. The Agreement on the Conservation of Small Cetaceans of the Baltic and North Seas (ASCOBANS) was adopted alongside ACCOBAMS to establish a special protection area for Europe's increasingly threatened cetaceans. The Agreement on the Conservation of Seals in the Wadden Sea, enforced in 1991, prohibits the killing or harassment of seals in the Wadden Sea, specifically targeting the harbor seal population.

In 1982, the United Nations Convention on the Law of the Sea (LOSC) adopted a pollution prevention approach to conservation, which many other conventions at the time also adopted.

As Food

Pilot whale meat (bottom), blubber (middle) and dried fish (left) with potatoes, Faroe Islands

For thousands of years, indigenous peoples of the Arctic have depended on whale meat. The meat is harvested from legal, non-commercial hunts that occur twice a year in the spring and autumn. The meat is stored and eaten throughout the winter. The skin and blubber (muktuk) taken from the bowhead, beluga, or narwhal is also valued, and is eaten raw or cooked. Whaling has also been practiced in the Faroe Islands in the North Atlantic since about the time of the first Norse settlements on the islands. Around 1000 Long-finned pilot whales are still killed annually, mainly during the summer. Today, dolphin meat is consumed in a small number of countries world-wide, which include Japan and Peru (where it is referred to as *chancho marino*, or "sea pork"). In some parts of the world, such as Taiji, Japan and the Faroe Islands, dolphins are traditionally considered food, and are killed in harpoon or drive hunts.

There have been human health concerns associated with the consumption of dolphin meat in Japan after tests showed that dolphin meat contained high levels of methylmercury. There are no known cases of mercury poisoning as a result of consuming dolphin meat, though the government continues to monitor people in areas where dolphin meat consumption is high. The Japanese government recommends that children and pregnant women avoid eating dolphin meat on a regular basis. Similar concerns exist with the consumption of dolphin meat in the Faroe Islands, where prenatal exposure to methylmercury and PCBs primarily from the consumption of pilot whale meat has resulted in neuropsychological deficits amongst children.

The Faroe Islands population was exposed to methylmercury largely from contaminated pilot whale meat, which contained very high levels of about 2 mg methylmercury/kg. However, the Faroe Islands populations also eat significant amounts of fish. The study of about 900 Faroese children showed that prenatal exposure to methylmercury resulted in neuropsychological deficits at 7 years of age

— World Health Organization

Ringed seals were once the main food staple for the Inuit. They are still an important food source for the people of Nunavut and are also hunted and eaten in Alaska. Seal meat is an important source of food for residents of small coastal communities. The seal blubber is used to make seal oil, which is marketed as a fish oil supplement. In 2001, two percent of Canada's raw seal oil was processed and sold in Canadian health stores.

In Captivity

Cetaceans

Performing killer whale at SeaWorld San Diego, 2009

Various species of dolphins are kept in captivity. These small cetaceans are more often than not kept in theme parks and dolphinariums, such as SeaWorld. Bottlenose dolphins are the most common species of dolphin kept in dolphinariums as they are relatively easy to train and have a long lifespan in captivity. Hundreds of bottlenose dolphins live in captivity across the world, though exact numbers are hard to determine. The dolphin "smile" makes them popular attractions, as this is a welcoming facial expression in humans; however the smile is due to a lack of facial muscles and subsequent lack of facial expressions.

Organizations such as World Animal Protection and the Whale and Dolphin Conservation Society campaign against the practice of keeping cetaceans, particularly killer whales, in captivity. In captivity, they often develop pathologies, such as the dorsal fin collapse seen in 60–90% of male killer whales. Captives have vastly reduced life expectancies, on average only living into their 20s. In the wild, females who survive infancy live 46 years on average, and up to 70–80 years in rare cases. Wild males who survive infancy live 31 years on average, and up to 50–60 years. Captivity usually bears little resemblance to wild habitat, and captive whales' social groups are foreign to those found in the wild. Captive life is also stressful due the requirement to perform circus tricks that are not part of wild killer whale behavior, as well as restricting pool size. Wild killer whales may travel up to 100 miles (160 km) in a day, and critics say the animals are too big and intelligent to be suitable for captivity. Captives occasionally act aggressively towards themselves, their tankmates, or humans, which critics say is a result of stress. Dolphins are often trained to do several anthropomorphic behaviors, including waving and kissing—behaviors wild dolphins would rarely do.

Pinnipeds

A sea lion trained to balance a ball on its nose

The large size and playfulness of pinnipeds make them popular attractions. Some exhibits have rocky backgrounds with artificial haul-out sites and a pool, while others have pens with small rocky, elevated shelters where the animals can dive into their pools. More elaborate exhibits contain deep pools that can be viewed underwater with rock-mimicking cement as haul-out areas. The most common pinniped species kept in captivity is the California sea lion as it is abundant and easy to train. These animals are used to perform tricks and entertain visitors. Other species popularly kept in captivity include the grey seal and harbor seal. Larger animals like walruses and Steller sea lions are much less common. Pinnipeds are popular attractions because they are "disneyfied", and, consequently, people often anthropomorphize them with a curious, funny, or playful nature.

Some organizations, such as the Humane Society of the United States and World Animal Protection, object to keeping pinnipeds and other marine mammals in captivity. They state that the exhibits could not be large enough to house animals that have evolved to be migratory, and a pool could never replace the size and biodiversity of the ocean. They also oppose using sea lions for entertainment, claiming the tricks performed are "exaggerated variations of their natural behaviors" and distract the audience from the animal's unnatural environment.

Others

Sea otters can do well in captivity, and are featured in over 40 public aquariums and zoos. The Seattle Aquarium became the first institution to raise sea otters from conception to adulthood with the birth of Tichuk in 1979, followed by three more pups in the early 1980s. In 2007, a YouTube video of two cute sea otters holding paws drew 1.5 million viewers in two weeks, and had over 20 million views as of January 2015. Filmed five years previously at the Vancouver Aquarium, it was YouTube's most popular animal video at the time, although it has since been surpassed. Otters are often viewed as having a "happy family life", but this is an anthropomorphism.

The oldest manatee in captivity is Snooty, at the South Florida Museum's Parker Manatee Aquarium in Bradenton, Florida. Born at the Miami Aquarium and Tackle Company on July 21, 1948, Snooty was one of the first recorded captive manatee births. He was raised entirely in captivity, and will never be released into the wild. Manatees can also be viewed in a number of European zoos, such as the Tierpark in Berlin, the Nuremberg Zoo, in ZooParc de Beauval in France, and in the Aquarium of Genoa in Italy. The River Safari at Singapore features seven of them.

Military

A dolphin wearing a locating pinger, performing mine clearance work in the Iraq War

Bottlenose dolphins and California sea lions were used in the United States Navy Marine Mammal Program (NMMP) to detect mines, protect ships from enemy soldiers, and recover objects. The Navy has never trained attack dolphins, as they would not be able to discern allied soldiers from enemy soldiers. There were five marine mammal teams, each purposed for one of the three tasks: MK4 (dolphins), MK5 (sea lions), MK6 (dolphins and sea lions), MK7 (dolphins), and MK8 (dolphins); MK is short for mark. The dolphin teams were trained to detect and mark mines either attached to the seafloor or floating in the water column, because dolphins can use their echolocative abilities to detect mines. The sea lion team retrieved test equipment such as fake mines or bombs dropped from planes usually out of reach of divers who would have to make multiple dives. MK6 protects harbors and ships from enemy divers, and was operational in the Gulf War and Vietnam War. The dolphins would swim up behind enemy divers and attach a buoy to their air tank, so that they would float to the surface and alert nearby Navy personnel. Sea lions would hand-cuff the enemy, and try to outmaneuver their counter-attacks.

The use of marine mammals by the Navy, even in accordance with the Navy's policy, continues to meet opposition. The Navy's policy says that only positive reinforcement is to be used while training the military dolphins, and that they be cared for in accordance with accepted standards in animal care. The inevitable stresses involved in training are topics of controversy, as their treatment is unlike the animals' natural lifestyle, especially towards their confined spaces when not training. There is also controversy over the use of muzzles and other inhibitors, which prevent the dolphins from foraging for food while working. The Navy states that this is to prevent them from ingesting harmful objects, but conservation activists say this is done to reinforce the trainers' control over the dolphins, who hand out food rewards. The means of transportation is also an issue for conservation activists, since they are hauled in dry carriers, and switching tanks and introducing the dolphin to new dolphins is potentially dangerous as they are territorial.

References

- Madigan M; Martinko J, eds. (2006). Brock Biology of Microorganisms (13th ed.). Pearson Education. p. 1096. ISBN 0-321-73551-X.

- Mahy WJ & Van Regenmortel MHV (eds). Desk Encyclopedia of General Virology. Oxford: Academic Press; 2009. ISBN 0-12-375146-2.

- Moyle, PB and Cech, JJ (2004) Fishes, An Introduction to Ichthyology. 5th Ed, Benjamin Cummings. ISBN 978-0-13-100847-2

- Madigan M; Martinko J, eds. (2006). Brock Biology of Microorganisms (13th ed.). Pearson Education. p. 1096. ISBN 0-321-73551-X.

- Jones, E. B. Gareth; Pang, Ka-Lai (2012-08-31). Marine Fungi: and Fungal-like Organisms. Walter de Gruyter. ISBN 9783110264067.

- Berta, A; Sumich, J. L. (1999). "Exploitation and conservation". Marine Mammals: Evolutionary Biology. San Diego: Academic Press. ISBN 978-0-12-093225-2. OCLC 42467530.

- Riedman, M. (1990). The Pinnipeds: Seals, Sea Lions, and Walruses. Los Angeles: University of California Press. ISBN 978-0-520-06497-3. OCLC 19511610.

- Whitehead, H. (2003). Sperm Whales: Social Evolution in the Ocean. Chicago: University of Chicago Press. p. 79. ISBN 978-0-226-89518-5. OCLC 51242162.

- Silverstein, Alvin; Silverstein, Virginia; Silverstein, Robert (1995). The Sea Otter. Brookfield, Connecticut: The Millbrook Press, Inc. p. 19. ISBN 978-1-56294-418-6. OCLC 30436543.

- Kenyon, Karl W. (1975). The Sea Otter in the Eastern Pacific Ocean. New York: Dover Publications. ISBN 978-0-486-21346-0. OCLC 1504461.

- Stirling, Ian (1988). "Distribution and Abundance". Polar Bears. Ann Arbor: University of Michigan Press. ISBN 0-472-10100-5.

- Lockyer, C. J. H.; Brown, S. G. (1981). "The Migration of Whales". In Aidley, D. Animal Migration. CUP Archive. p. 111. ISBN 978-0-521-23274-6.

- Berta, A.; Sumich, J. L.; Kovacs, K. M. (2015). Marine Mammals: Evolutionary Biology. London: Academic Press. p. 430. ISBN 978-0-12-397002-2. OCLC 905649783.

- VanBlaricom, Glenn R. (2001). Sea Otters. Stillwater, MN: Voyageur Press Inc. pp. 22, 33, 69. ISBN 978-0-89658-562-1. OCLC 46393741.

- Lavinge, D. M.; Kovacs, K. M.; Bonner, W. N. (2001). "Seals and Sea lions". In MacDonald, D. The Encyclopedia of Mammals (2nd ed.). Oxford University Press. pp. 147–55. ISBN 978-0-7607-1969-5. OCLC 48048972.

- Marsh, Helene; O'Shea, Thomas J.; Reynolds III, John E. (2012). Ecology and Conservation of the Sirenia: Dugongs and Manatees. Cambridge: Cambridge University Press. p. 112. ISBN 978-0-521-88828-8. OCLC 773872519.

- Marsh, Helene. "Dugongidae". Fauna of Australia. 1. Canberra: Australian Government Public Service. ISBN 978-0-644-06056-1. OCLC 27492815.

- Stirling, Ian; Guravich, Dan (1988). Polar Bears. Ann Arbor, MI: University of Michigan Press. pp. 27–28. ISBN 978-0-472-10100-9. OCLC 757032303.

- Heimlich, Sara; Boran, James (2001). Killer Whales. Stillwater, Minnesota: Voyageur Press. ISBN 978-0-89658-545-4. OCLC 46973039.

- Haycox, Stephen W. (2002). Alaska: An American Colony. University of Washington Press. pp. 53–58. ISBN 978-0-295-98249-6. OCLC 49225731.

- Riedman, M. (1990). The Pinnipeds: Seals, Sea Lions, and Walruses. San Francisco: University of California Press. ISBN 978-0-520-06497-3. OCLC 19511610.

- Ora, Nilufer (2013). Regional Co-operation and Protection of the Marine Environment Under International Law. Leiden, Netherlands: Koninklijke Brill. pp. 131–137. ISBN 978-90-04-25085-7.

- Braathen, Jonette N. (1998). International Co-operation on Fisheries and Environment. TemaNord. Copenhagen: Nordic Council of Ministers. p. 45. ISBN 978-92-893-0198-5.

- White, Thomas (2007). In Defense of Dolphins: The New Moral Frontier. Malden, MA: Blackwell Publishing. p. 17. ISBN 978-1-4051-5779-7. OCLC 122974162.

- Eglan, Jared (2015). Beasts of War: The Militarization of Animals. Lulu.com. pp. 126–128. ISBN 978-1-329-51613-7.

Permissions

Index

www.ingramcontent.com/pod-product-compliance
Lightning Source LLC
Chambersburg PA
CBHW061935190326
41458CB00009B/2745